Microbiology Research Advances

www.novapublishers.com

Microbiology Research Advances

Endophytes: Types, Potential Uses and Mechanism(s) of Action
Pragya Tiwari, PhD (Editor)
2022. ISBN: 979-8-88697-045-6 (Hardcover)
2022. ISBN: 979-8-88697-205-4 (eBook)

Endotoxins and Their Importance
Arif Pandit, PhD and R. S. Sethi, PhD (Editors)
2022. ISBN: 978-1-68507-839-3 (Softcover)
2022. ISBN: 978-1-68507-913-0 (eBook)

Understanding Antibiofilm Activity
Iram Liaqat, PhD (Editor)
2022. ISBN: 978-1-68507-927-7 (Hardcover)
2022. ISBN: 979-8-88697-005-0 (eBook)

Bacteriophages: Interaction, Diversity and Applications
Prasanth Manohar, PhD (Editor)
2022. ISBN: 978-1-68507-860-7 (Hardcover)
2022. ISBN: 978-1-68507-958-1 (eBook)

Endophytic Fungi:
Biodiversity, Antimicrobial Activity and Ecological Implications
Qiang-Sheng Wu, PhD, Ying-Ning Zou
and Yong-Jie Xu, PhD
(Editors)
2021. ISBN: 978-1-68507-354-1 (Softcover)
2021. ISBN: 978-1-68507-442-5 (eBook)

More information about this series can be found at
https://novapublishers.com/product-category/series/microbiology-research-advances/

Kaley Rutherford
Editor

Catalase and Its Applications

Copyright © 2022 by Nova Science Publishers, Inc.
DOI: https://doi.org/10.52305/YAEW1520

All rights reserved. No part of this book may be reproduced, stored in a retrieval system or transmitted in any form or by any means: electronic, electrostatic, magnetic, tape, mechanical photocopying, recording or otherwise without the written permission of the Publisher.

We have partnered with Copyright Clearance Center to make it easy for you to obtain permissions to reuse content from this publication. Simply navigate to this publication's page on Nova's website and locate the "Get Permission" button below the title description. This button is linked directly to the title's permission page on copyright.com. Alternatively, you can visit copyright.com and search by title, ISBN, or ISSN.

For further questions about using the service on copyright.com, please contact:
Copyright Clearance Center
Phone: +1-(978) 750-8400 Fax: +1-(978) 750-4470 E-mail: info@copyright.com.

NOTICE TO THE READER

The Publisher has taken reasonable care in the preparation of this book, but makes no expressed or implied warranty of any kind and assumes no responsibility for any errors or omissions. No liability is assumed for incidental or consequential damages in connection with or arising out of information contained in this book. The Publisher shall not be liable for any special, consequential, or exemplary damages resulting, in whole or in part, from the readers' use of, or reliance upon, this material. Any parts of this book based on government reports are so indicated and copyright is claimed for those parts to the extent applicable to compilations of such works.

Independent verification should be sought for any data, advice or recommendations contained in this book. In addition, no responsibility is assumed by the Publisher for any injury and/or damage to persons or property arising from any methods, products, instructions, ideas or otherwise contained in this publication.

This publication is designed to provide accurate and authoritative information with regard to the subject matter covered herein. It is sold with the clear understanding that the Publisher is not engaged in rendering legal or any other professional services. If legal or any other expert assistance is required, the services of a competent person should be sought. FROM A DECLARATION OF PARTICIPANTS JOINTLY ADOPTED BY A COMMITTEE OF THE AMERICAN BAR ASSOCIATION AND A COMMITTEE OF PUBLISHERS.

Additional color graphics may be available in the e-book version of this book.

Library of Congress Cataloging-in-Publication Data

ISBN: 979-8-88697-421-8

Published by Nova Science Publishers, Inc. † New York

Contents

Preface

This book is a collection of five chapters discussing catalase and its various applications in different fields. Chapter One discusses the cardiovascular effects of a catalase blockade of H_2O_2 in normotensive and hypertensive rats. Chapter Two reviews daily and seasonal changes in activity of catalase in littoral macroalgae on the Barents Sea, as well as the influence of abiotic factors in the aforementioned. Chapter Three is an overview of plant catalases under abiotic stress. Chapter Four analyses aspects of the blood catalase in the human body, and finally, Chapter Five discusses the inhibitory effect of drugs on catalase activity.

Chapter 1 - Reactive oxygen species (ROS) like superoxide anion ($O_2^{\cdot -}$), hydroxyl radical (HO•), and hydrogen peroxide (H_2O_2) are highly reactive chemical structures derived from molecular oxygen. Although ROS may accumulate in cells causing deleterious effects, at low concentrations they have an important role in the control of cell signaling modulating the cellular action of traditional neurotransmitters or hormones such as dopamine, GABA, glutamate, or angiotensin II (ANG II). Brain adenoviral-mediated overexpression of superoxide dismutase (SOD), the enzyme that dismutates $O_2^{\cdot -}$ to H_2O_2, abolished the pressor and dipsogenic responses to intracerebroventricular (icv) ANG II, suggesting that icv ANG II responses depend on $O_2^{\cdot -}$ formation. Another possibility is that increased H_2O_2 production activates mechanisms that reduce the effects of ANG II. The icv injection of H_2O_2 or the blockade of catalase (an enzyme that degrades H_2O_2 into H_2O and O_2) with icv injection of 3-amino-1,2,4-triazole (ATZ) reduced the pressor effects of ANG II also injected icv, which suggests that exogenous or endogenous H_2O_2 activates mechanisms that inhibits the cardiovascular effects of ANG II (Lauar et al. 2010). The icv or intravenous (iv) treatment with ATZ or icv H_2O_2 also reduced the pressor response induced by carbachol (cholinergic muscarinic agonist) icv, suggesting that endogenous or exogenous H_2O_2 acting centrally probably decreases AVP release and/or sympathetic activation produced by carbachol. The pre-treatment of the medial septal area (MSA) with H2O2 reduced water intake, natriuresis, kaliuresis, antidiuresis, and the pressor response to the injection of carbachol into the same area, an effect suggested depending at the least in part on reduction of vasopressin release due to KATP channel activation (Melo et al. 2015). The icv injection of H2O2 or ATZ reduced the pressor responses to icv ANG II also in spontaneously hypertensive rats (SHRs) and in 2 kidneys, 1 clip (2K1C) hypertensive rats. The iv injection of ATZ alone or combined with icv injection of H2O2 also reduced the pressor response induced by icv injection of ANG II in SHRs and 2K1C hypertensive rats. In addition, baseline arterial pressure was also acutely reduced in 2K1C hypertensive rats treated with icv injection of H2O2, iv injection of ATZ, or a combination of both, and in SHRs treated with icv injection of H2O2 alone or in combination with iv injection of ATZ. These results suggest that the increase of endogenous H2O2 produced by the blockade of catalase with ATZ induces anti-hypertensive effects due to the impairment of central pressor mechanisms activated by ANG II in SHRs and 2K1C hypertensive rats.

Chapter 2 - The enzyme catalase is an important component of the antioxidant system of algae. Its activity depends on their physiological state and on environmental factors. The catalase activity depends on the age of the thallus: in brown macroalgae *Fucus vesiculosus* the maximum is peculiar for the young part of the thallus (apical and middle), minimal observe in basal part. The presence of strong (2-4 times) fluctuations in catalase activity in *F. vesiculosus* during the day, which are synchronized with the tidal cycle, is shown. The main increases occur in the middle of low tide, as well as during periods when there is a change of habitat (transition from water to air and back). Especially significant changes are observed with a significant temperature/light gradient. In addition to daily changes in catalase activity, seasonal changes are peculiar also. During the year, the minimum catalase activity in *F. vesiculosus*, is observed in April, the maximum in January. During the summer and autumn, the activity of catalase practically does not change. In the red alga *Palmaria palmata*, the maximum catalase activity is observed in winter (December-February), the minimum in summer (July). Both types of algae have a similar tendency in changing catalase activity by seasons: two periods can be distinguished, spring and autumn, associated with the adaptation of biochemical processes to changes in ambient temperature. In the spring, when the temperature and PAR increase, catalase inactivation occurs. In autumn, with a decrease in ambient temperature and PAR, catalase activity increases. During the period of decrease in catalase activity, its role can be performed by a complex of peroxidases. Catalase activity increases in algae (*F. vesiculosus*) in the presence of an petroleum products in the environment. Experimental studies have shown that catalase, along with other AOS enzymes, is activated in the first hours of the action of the toxicant (diesel fuel), and remains at a high level until the algae adapt. In general, the enzyme catalase is a sensitive component of the AOS of algae, which quickly reacts to changes in environmental factors and participates in increasing the adaptation of the organism.

Chapter 3 - Hydrogen peroxide (H2O2) is produced in plants under normal and stressed conditions. At low concentration, this signal molecule is involved in different biological/physiological processes such as growth and development, cell cycle, photosynthetic functions, and plant responses to biotic and abiotic stresses. Interestingly, accumulation of H2O2 can cause Oxidative stress and then eventual cell death. Thus, plants developed a preferment antioxidant enzymatic system to defend against threats. Enzymatic antioxidant defense occurs as a series of redox reactions for ROS elimination. Different enzymes are implicated in this process such as catalase (CAT), ascorbate peroxidases (APX), and superoxide dismutase (SOD). Catalase is a crucial enzyme in antioxidant defense system protecting eukaryotes from oxidative stress. Those proteins are present in almost all living organisms and play important roles in controlling plant response to biotic and abiotic stresses by catalyzing the decomposition of H2O2. Based on recent reports, this chapter highlights the role of catalase in plant defense against accumulation of H2O2 to regulate plant response to different abiotic stresses.

Chapter 4 - The catalase enzyme is one of the earliest known and well studied enzymes. In the decades following its discovery in 1819, the enzyme catalase has been studied and detected in many types of living organisms, organs and tissues. Its substrate hydrogen peroxide was discovered in 1818. In the body, the catalase enzyme is responsible for eliminating hydrogen peroxide in toxic concentrations. The enzyme is encoded by a single gene (33114 Kb, 13 exons, 12 introns). Several polymorphisms and mutations have been identified for the catalase gene. Red blood cells contain 99.9% of blood catalase. The activity of the blood catalase can be easily determined by a widely used spectrophotometric method. It is a combination of optimized enzymatic conditions and the spectrophotometric assay of hydrogen peroxide based on formation of its stable complex with ammonium molybdate which could be measured at 405 nm. Acatalasemia occurs when the activity of the enzyme blood catalase is less than 10%. Hypocatalasemia is when blood cell activity is below 50% of the average. The authors have reported patients with acatalasemia and hypocatalasemia in diseases such as diabetes mellitus, microcyter anemia, vitiligo, and thalassemia. We've identified mutations and polymorphisms in the catalase gene that may be responsible for the decrease in enzyme activity. Mutations and polymorphisms are in exons and introns. The effects of the polymorphisms in the introns are currently unknown. According to the exon mutations of catalase gene eleven types of acatalasemia were established in Hungary. Enzyme catalase plays an important role in defense against the toxic effects of free oxygen species by the removal of hydrogen peroxide. Hydrogen peroxide may be formed in several pathologic conditions such as inflammation, cancer, erythrocyte metabolism, homeostasis and platelet activation. In this chapter the authors are discussing: the catalase enzyme (discovery, history, the relationship with hydrogen peroxide and reactive oxygen species, the blood and erythrocyte catalase), the catalase gene (its deficiency acatalasemia and hypocatalasemia, its mutations), the blood catalase in various diseases (diabetes mellitus, anemia, thalassemia, vitiligo) and the possible future of catalase research.

Chapter 5 - Catalase belongs to the group of oxidoreductases, and as an antioxidant enzyme, it plays a major role in eliminating hydrogen peroxide, a by-product of many normal cellular reactions. Reduced catalase activity increases oxidative stress in the body. The link between oxidative stress and many diseases suggests that each endogenous catalase inhibitor has the potential to contribute to the occurrence of certain pathological conditions. A non-functional antioxidant system with reduced catalase activity in the cell can lead to pathological conditions such as certain types of cancer, Alzheimer's disease, diabetes, including inflammatory processes. Certain drugs given for therapeutic purposes can reduce the activity of catalase in the cell, so in this chapter, the authors will talk about the kinetics and mechanism of catalase inhibition. Drugs exert various effects on catalase, acting as competitive, non-competitive or a competitive enzyme inhibitors. In this process, the drug molecule undergoes a chemical transformation and becomes a suitable product that blocks the normal flow of the metabolic pathway. Inhibitors specifically bind to enzymes and modify the enzymatic reaction rate. The same substance in relation to the substrate concentration can be both an activator and an inhibitor of the test enzyme. Due to structural changes resulting from substrate-inhibitor binding, partial inhibitors may also be allosteric inhibitors of enzyme activity. By determining the kinetic constants of the Michaelis-Menten constant, the maximum rate of the enzymatic reaction, the catalytic constant and the inhibition constant, information can be obtained on the enzyme catalysis efficiency and the mechanism of catalase inhibition itself.

Chapter 1

Cardiovascular Effects of Catalase Blockade/H_2O_2 in Normotensive and Hypertensive Rats

Mariana R. Lauar, Débora S. A. Colombari, Eduardo Colombari, Patrícia M. De Paula, Laurival A. De Luca Jr, Carina A. F. Andrade and José V. Menani*

Department of Physiology and Pathology, Dentistry School, UNESP, Araraquara, Brazil

Abstract

Reactive oxygen species (ROS) like superoxide anion ($O_2^{\cdot -}$), hydroxyl radical ($HO^\bullet$), and hydrogen peroxide (H_2O_2) are *highly reactive chemical structures* derived from molecular oxygen. Although ROS may accumulate in cells causing deleterious effects, at low concentrations they have an important role in the control of cell signaling modulating the cellular action of traditional neurotransmitters or hormones such as dopamine, GABA, glutamate, or angiotensin II (ANG II). Brain adenoviral-mediated overexpression of superoxide dismutase (SOD), the enzyme that dismutates $O_2^{\cdot -}$ to H_2O_2, abolished the pressor and dipsogenic responses to intracerebroventricular (icv) ANG II, suggesting that icv ANG II responses depend on $O_2^{\cdot -}$ formation. Another possibility is that increased H_2O_2 production activates mechanisms that reduce the effects of ANG II. The icv injection of H_2O_2 or the blockade of catalase (an enzyme that degrades H_2O_2 into H_2O and O_2) with icv injection of 3-

* Corresponding Author's Email: jv.menani@unesp.br.

In: Catalase and Its Applications
Editor: Kaley Rutherford
ISBN: 979-8-88697-421-8
© 2022 Nova Science Publishers, Inc.

amino-1,2,4-triazole (ATZ) reduced the pressor effects of ANG II also injected icv, which suggests that exogenous or endogenous H_2O_2 activates mechanisms that inhibits the cardiovascular effects of ANG II (Lauar et al. 2010). The icv or intravenous (iv) treatment with ATZ or icv H_2O_2 also reduced the pressor response induced by carbachol (cholinergic muscarinic agonist) icv, suggesting that endogenous or exogenous H_2O_2 acting centrally probably decreases AVP release and/or sympathetic activation produced by carbachol (Lauar et al. 2019). The pre-treatment of the medial septal area (MSA) with H_2O_2 reduced water intake, natriuresis, kaliuresis, antidiuresis, and the pressor response to the injection of carbachol into the same area, an effect suggested depending at the least in part on reduction of vasopressin release due to K_{ATP} channel activation (Melo et al. 2015). The icv injection of H_2O_2 or ATZ reduced the pressor responses to icv ANG II also in spontaneously hypertensive rats (SHRs) and in 2 kidneys, 1 clip (2K1C) hypertensive rats. The iv injection of ATZ alone or combined with icv injection of H_2O_2 also reduced the pressor response induced by icv injection of ANG II in SHRs and 2K1C hypertensive rats. In addition, baseline arterial pressure was also acutely reduced in 2K1C hypertensive rats treated with icv injection of H_2O_2, iv injection of ATZ, or a combination of both, and in SHRs treated with icv injection of H_2O_2 alone or in combination with iv injection of ATZ. These results suggest that the increase of endogenous H_2O_2 produced by the blockade of catalase with ATZ induces anti-hypertensive effects due to the impairment of central pressor mechanisms activated by ANG II in SHRs and 2K1C hypertensive rats (Lauar et al. 2020).

Keywords: arterial pressure, angiotensin II, catalase inhibitor, hydrogen peroxide, autonomic modulation, hypertension

Introduction

The reactive oxygen species (ROS) like superoxide anion ($O_2^{\cdot -}$) and hydrogen peroxide (H_2O_2) are produced endogenously and may participate in intra- and extracellular signaling, including mediation of angiotensin II (ANG II) responses. Evidence indicates that the activation of ANG II type 1 (AT1) receptor enhances nicotinamide adenine dinucleotide phosphate (NAD(P)H) oxidase activity, leading to the formation of ROS like $O_2^{\cdot -}$ and H_2O_2 (Griendling et al. 1994; Berry et al. 2000; Zimmerman, Lazartigues, et al. 2004). An enzyme involved in the oxidative metabolism widely distributed in the CNS is the superoxide dismutase (SOD) that catalyzes the dismutation of

$O_2^{\bullet-}$ producing H_2O_2 which can be converted by the catalase enzyme to oxygen and water. These ROS produced endogenously act as a cellular signaling molecule to regulate biological function or as neuromodulators affecting neurotransmission and neuronal firing (Aizenman, Lipton, and Loring 1989; Volterra et al. 1994; Zoccarato, Valente, and Alexandre 1995; Adler et al. 1999; Chen, Avshalumov, and Rice 2001; Zimmerman et al. 2002; Zimmerman and Davisson 2004; Avshalumov et al. 2005; Bao et al. 2009). Previous studies have suggested that increased degradation of $O_2^{\bullet-}$ produced by the action of ANG II on the AT1 receptor by central adenovirus-mediated overexpression of SOD abolishes the pressor response to central injections of ANG II, suggesting that $O_2^{\bullet-}$ is part of the signaling mechanisms activated by ANG II centrally (Zimmerman et al. 2002; Zimmerman and Davisson 2004). However, H_2O_2 levels might also be influenced by changes in SOD activity (Teixeira, Schumacher, and Meneghini 1998; Gardner, Salvador, and Moradas-Ferreira 2002; Chan et al. 2006; Kowald, Lehrach, and Klipp 2006) and, therefore, central SOD overexpression might also reduce ANG II-induced responses due to increases in H_2O_2 levels. Here we demonstrate how the inhibition of the catalase enzyme leading to an increase in H_2O_2 levels might modulate ANG II-induced pressor responses.

Cardiovascular Control by the Renin Angiotensin System

ANG II, the main peptide of the renin angiotensin system (RAS), has the formation stimulated by a decrease in renal perfusion pressure, reduced sodium ion concentration in the macula densa cells in the juxtaglomerular apparatus of the kidneys, or stimulation of renal β1 adrenergic receptors. These stimuli induce the release of renin which leads to the formation of ANG I which is transformed into ANG II by the action of angiotensin converting enzyme (ACE) (Cowley and Guyton 1972; Hackenthal et al. 1990; Hodge, Lowe, and Vane 1966). ANG II acts in different tissues, such as the adrenal gland, stimulating the synthesis and release of aldosterone, which increases sodium reabsorption; in vascular beds leading to vasoconstriction, increasing peripheral resistance; in the heart leading to increased force of contraction; in the kidneys reducing renal blood flow and increasing sodium reabsorption; in the central nervous system (CNS) stimulating thirst, sodium appetite, vasopressin release, and increased sympathetic activity; and interacting with growth factors favoring cell proliferation (Blair-West et al. 1998; Fitzsimons 1998; Liang and Gavras 1978; Navar et al. 1998; Santos RAS & Sampaio WO

2002). These effects are due to the actions of ANG II, predominantly on AT1 receptors, which are receptors coupled to the Gq protein, responsible for stimulating the activity of phospholipase C, determining the hydrolysis of the phosphatidylinositol polyphosphates of the plasma membrane to form inositol polyphosphates. The main isomer of inositol phosphate, (1, 4, 5) inositol triphosphate – IP3 promotes the release of Ca^{2+} from intracellular stores by binding to an endoplasmic reticulum membrane receptor. The second product of phospholipase action, diacylglycerol (DAG), activates protein kinase C (in conjunction with calcium/calmodulin), which controls several cellular functions through the phosphorylation of several proteins, culminating in the opening of ion channels that will lead to a cellular depolarization (Phillips and Sumners 1998; M. Zhu et al. 1999).

More recently, it was described that the components of the RAS, including precursors and enzymes necessary for the production of ANG II, were identified in the CNS, suggesting the participation of ANG II as a neuromodulator and/or neurotransmitter (de Kloet et al. 2015; Grobe, Xu, and Sigmund 2008; Wright and Harding 2013). Studies have shown that astrocytes secrete angiotensinogen in the interstitial space and in the extracellular fluid (Deschepper, Bouhnik, and Ganong 1986; Stornetta et al. 1988) and that neurons in the paraventricular nucleus of the hypothalamus (PVN) are capable of producing angiotensinogen (Aronsson et al. 1988). Furthermore, studies have shown that renin (Ganten et al. 1971) and ACE (Correa, Plunkett, and Saavedra 1986; Dzau et al. 1986; Saavedra, Fernandez-Pardal, and Chevillard 1982) are also present in the CNS. AT1 receptors are located diffusely throughout the brain, however, are predominantly present in neurons and glial cells of the circumventricular organs (CVOs). The CVOs directly activated by ANG II acting centrally are the subfornical organ (SFO) and the organum vasculosum of the lamina terminalis (OVLT). From these regions, facilitatory signals can reach the PVN and the rostroventrolateral region of the medulla (RVLM) to increase sympathetic activity and to the PVN and supraotic nucleus of the hypothalamus (SON) to increase vasopressin secretion (Hoffman et al. 1977; Buggy et al. 1977; Johnson 1985; Johnson, Hoffman, and Buggy 1978; Mahon et al. 1995) which will culminate in an increase in blood pressure (BP). In addition, these hypothalamic areas send projections to caudal areas such as the parabrachial nucleus (PBN) and the nucleus of the solitary tract (NTS) (Phillips and Sumners 1998). The NTS is the first synaptic station where signals from visceral afferents converge, such as arterial baroreceptor signals (Palkovits and Záborszky 1977; Torvik 1956). Furthermore, it has been demonstrated that ANG II acting on AT1 receptors in the NTS promotes an attenuation in baroreflex function, an important mechanism for moment-to-moment regulation of BP (Matsumura, Averill, and Ferrario 1998; Michelini and Bonagamba 1990; Paton and Kasparov 1999).

Biological Action of ROS

ROS, known for their microbicidal properties, are chemical species endogenously produced by the body such as free radicals, such as the superoxide anion ($O_2^{\bullet-}$) and the hydroxyl radical ($HO^{\bullet}$) in addition to non-radical species like hydrogen peroxide (H_2O_2) (Rhee et al. 2003; Weinberg 1990). The formation of ROS is initiated by the activation of the NADPH oxidase enzyme that leads to the formation of $O_2^{\bullet-}$ by the incomplete reduction of O_2 [reviewed in (Cohen 1994)]. The $O_2^{\bullet-}$ is dismutated by the enzyme superoxide dismutase (SOD) in H_2O_2. H_2O_2 is then degraded by the catalase enzyme into H_2O and O_2 or is converted to $HO^{\bullet}$ by contact with the ferrous ion (McCord and Fridovich 1969).

NADPH oxidase is an enzymatic complex formed by the membrane subunit gp91phox and p22phox, by the cytoplasmic subunits p^{47phox}, $p^{40phox,}$ and $p^{67phox,}$ and by the G proteins Rac and Rap1a (Lassègue and Clempus 2003). Among these subunits, it is important to highlight the catalytic site gp^{91phox}, which has seven different isoforms (NOX1-NOX5 and DUOX1/DUOX2) [review in (Álvarez et al. 2013)]. The NOX2 and NOX4 isoforms are widely expressed in the CNS and are therefore the most studied.

The high oxidative property of ROS may cause damage to cells due to their destructive properties and is often associated with cell death and apoptosis (Irani 2000; Jiang et al. 2003). Despite their cytotoxic potential in high concentrations, studies have shown that these species, when in low concentrations, can participate in intra and intercellular signaling, acting together with classic chemical mediators such as ANG II, noradrenaline, dopamine, and glutamate (Rice 2011; Zimmerman and Davisson 2004).

Overexpression of SOD in the Brain on Angiotensin II Effects

ANG II acts centrally to produce pressor responses dependent on sympathetic activation and vasopressin secretion, as well as producing natriorexigenic and dipsogenic responses (Hoffman et al. 1977; Johnson, Hoffman, and Buggy

1978; Johnson 1985; Mahon et al. 1995; Fitzsimons 1998). Moreover, acting on AT1 receptors ANG II may produce ROS in response to the activation of the NADPH oxidase enzyme complex that generates $O_2^{\bullet-}$ (Lassègue and Clempus 2003) in the vascular wall and possibly in the mediation of the central effects of ANG II (Sun et al. 2005; Vaziri et al. 2002; G.-Q. Zhu et al. 2004; Zimmerman et al. 2002; Zimmerman and Davisson 2004).

Silencing of NOX2 and/or NOX4 in the SFO reduce the pressor response induced by icv ANG II injection (Peterson et al. 2009). Furthermore, in the CNS, the dipsogenic and cardiovascular actions of ANG II in mice have been reported to be dependent on increased $O_2^{\bullet-}$ production by CVO neurons (Zimmerman and Davisson 2004; Zimmerman et al. 2002). According to the proposed mechanism, a decrease in ANG II-induced $O_2^{\bullet-}$ formation by central adenovirus-mediated overexpression of SOD, almost abolishes the pressor and dipsogenic responses to central injections of ANG II, suggesting that $O_2^{\bullet-}$ is part of the signaling mechanisms activated by ANG II centrally (Zimmerman et al. 2002; Zimmerman and Davisson 2004; Zimmerman, Dunlay, et al. 2004). However, those studies did not take into account that although the increase in SOD activity decreases the availability of $O_2^{\bullet-}$ it can also increase the availability of H_2O_2 when the NADPH oxidase complex is triggered by ANG II. Therefore, the effects observed when the SOD activity increases may depend not only on the reduction in the availability of $O_2^{\bullet-}$, but also on the increase in the availability of H_2O_2.

The importance of $O_2^{\bullet-}$ as part of the signaling mechanisms activated by ANG II was tested by different studies, however, possible effects of other ROS, like H_2O_2, on ANG II-induced responses were poorly investigated. Although questions may exist about a correlation between changes in SOD activity and H_2O_2 levels (Teixeira, Schumacher, and Meneghini 1998; Gardner, Salvador, and Moradas-Ferreira 2002; Chan et al. 2006; Kowald, Lehrach, and Klipp 2006), SOD overexpression might also increase H_2O_2 levels, which in turn might affect the responses to ANG II acting centrally. In this case, $O_2^{\bullet-}$ produced by the action of ANG II on AT1 receptors is dismutated by SOD to H_2O_2, and H_2O_2 produced might counterbalance the effects of ANG II, limiting its responses. With SOD overexpression, not only $O_2^{\bullet-}$ levels reduced, but simultaneously H_2O_2 levels increase, and both together may account for the reduced ANG II responses produced by SOD overexpression.

Endogenously, H_2O_2 production may result from different sources (Maker et al. 1981; Zimmerman, Dunlay, et al. 2004; Peterson et al. 2009; Bao, Avshalumov, and Rice 2005; Bao et al. 2009). Independently from the source,

H_2O_2 might affect ANG II-induced responses through mechanisms that reduce neuronal excitability. H_2O_2 is a relatively stable and diffusible ROS that may act centrally through different mechanisms producing excitatory or inhibitory responses (Sorg et al. 1997; Volterra et al. 1994; Zoccarato, Valente, and Alexandre 1995; Zoccarato et al. 1999; Sah et al. 2002; Wehage et al. 2002; Bao, Avshalumov, and Rice 2005; Avshalumov et al. 2005; Takahashi, Mikami, and Yang 2007). H_2O_2 may act centrally blocking glutamate uptake by glial cells causing an increase in extracellular glutamate levels and enhancing neuronal excitability (Sorg et al. 1997; Volterra et al. 1994). In the opposite direction, H_2O_2 may inhibit glutamate and increase GABA release and action (Zoccarato, Valente, and Alexandre 1995; Zoccarato et al. 1999; Sah et al. 2002; Takahashi, Mikami, and Yang 2007). H_2O_2 may also act on ATP-sensitive potassium channels (KATP channels) causing neuronal hyperpolarization and reducing neuronal excitability (Bao, Avshalumov, and Rice 2005; Avshalumov et al. 2005).

In the present chapter, it is commented results and conclusions from studies that investigated the effects of exogenous H_2O_2 injected icv or the increase of endogenous H_2O_2 produced by icv or iv injections of the catalase inhibitor 3-amino-1,2,4-triazole (ATZ) (Aragon, Rogan, and Amit 1991; Heim, Appleman, and Pyfrom 1955; Nicholls 1962) on the pressor responses induced by icv ANG II or central cholinergic activation in normotensive rats. In addition, results from H_2O_2 icv and ATZ iv or icv on baseline arterial pressure and on the pressor response to ANG II icv in hypertensive rats are also commented.

Method

Animals and Treatments

Adult male normotensive Holtzman rats or male Holtzman rats that received a clip in the left renal artery to produce the 2 kidneys, one clip (2K1C) hypertension, and spontaneously hypertensive rats (SHRs) with stainless steel cannulas implanted in the lateral ventricle (LV) were used. A polyethylene tubing (PE-10 connected to a PE-50) was inserted into the abdominal aorta through the femoral artery for mean arterial pressure (MAP) and heart rate (HR) recordings. Another PE-10 connected to a PE-50 was inserted into the femoral vein for drug administration. It was injected hydrogen peroxide

(H_2O_2, 5 µmol into the LV), angiotensin II (50 ng into the LV), 3-amino-1,2,4-triazole (ATZ, 5 nmol into the LV, and 3.6 mmol/kg of body weight iv) and carbachol (4 nmol into the LV). The volume injected into the LV was 1 µl. The vehicle was PBS or saline. MAP and HR were recorded in freely moving rats. For the confirmation of the sites of injections, the brains were removed, fixed in 10% formalin, frozen, cut into 50 µm sections, and analyzed by light microscopy.

Result and Discussion

Responses to Central Angiotensinergic Stimulation in Normotensive Rats Treated with H_2O_2 or ATZ icv

H_2O_2 (5 µmol) injected icv in normotensive rats reduced the pressor response to icv ANG II (50 ng) by 63%, whereas H_2O_2 injected iv produced no change in the pressor response to iv ANG II (Table 1) (Lauar et al. 2010). H_2O_2 (5 µmol) injected iv or icv in normotensive rats produced no change in baseline MAP and HR (Table 1) (Lauar et al. 2010).

The treatment with ATZ (5 nmol/1 µl) icv in normotensive rats reduced the pressor response to ANG II icv by 78%, without changing baseline MAP or HR (Table 1) (Lauar et al. 2010). ANG II icv reduced HR only after the pre-treatment with ATZ (Lauar et al. 2010).

These results suggest that H_2O_2 acting centrally inhibits the pressor mechanisms activated by central angiotensinergic stimulation (Lauar et al. 2010).

Responses to Central Cholinergic Stimulation in Normotensive Rats Treated with H_2O_2 or ATZ icv

H_2O_2 injected icv in normotensive rats reduced the pressor response to icv carbachol (cholinergic agonist, 4 nmol/1 µl) by 52% (Table 1) (Lauar et al. 2019).

In normotensive rats, ATZ (5 nmol/1 µl) icv reduced by 57% the pressor response to carbachol icv (Table 1) (Lauar et al. 2019), without changing baseline MAP. ATZ and carbachol icv alone or combined produced no change in HR (Lauar et al. 2019).

These results suggest that H_2O_2 acting centrally inhibits the pressor mechanisms activated by central cholinergic stimulation (Lauar et al. 2019).

Table 1. Baseline arterial pressure and pressor responses to ANG II or carbachol icv or iv in normotensive rats treated with H_2O_2 or ATZ icv or iv

Treatment	Effects	% of change
H_2O_2 icv	No change of baseline MAP	-
H_2O_2 icv + ANG II icv	Reduction of the pressor response	63% reduction
H_2O_2 iv	No change of baseline MAP	-
H_2O_2 iv + ANG II iv	No change in the pressor response	-
H_2O_2 icv + carbachol icv	Reduction of the pressor response	52% reduction
ATZ icv	No change of baseline MAP	-
ATZ icv + ANG II icv	Reduction of the pressor response	78% reduction
ATZ icv + carbachol icv	Reduction of the pressor response	57% reduction
ATZ iv	No change of baseline MAP	-
ATZ iv + ANG II icv	Reduction of the pressor response	66% reduction
ATZ iv + carbachol icv	Reduction of the pressor response	32% reduction

H_2O_2, 5 μmol; ANG II, 50 ng; ATZ, 5 nmol icv and 3.6 mmol/kg of body weight iv; carbachol, 4 nmol.

Responses to Central Angiotensinergic or Cholinergic Stimulation in Normotensive Rats Treated with ATZ iv

In normotensive rats, iv injections of ATZ (3.6 mmol/kg of body weight) also reduced the pressor response to ANG II icv by 66%, without changing baseline MAP and HR (Table 1) (Lauar et al. 2010). ANG II icv alone or combined with ATZ produced no change in HR.

ATZ iv also reduced by 32% the pressor response to carbachol icv in normotensive rats (Table 1) (Lauar et al. 2019).

These results suggest that iv injections of ATZ also inhibit the pressor mechanisms activated by central angiotensinergic and cholinergic stimulation (Lauar et al. 2010; 2019).

Responses to Central Angiotensinergic Stimulation in 2K1C Hypertensive Rats Treated with H_2O_2 or ATZ icv

H_2O_2 injected icv in 2K1C hypertensive rats reduced by 80% the pressor response to icv ANG II (Table 2) (Lauar et al. 2020). H_2O_2 injected icv in

2K1C hypertensive rats slightly reduced baseline MAP acutely (around 10 mmHg in the next 20 min), without significant changes in baseline HR (Table 2) (Lauar et al. 2020).

Table 2. Baseline arterial pressure and pressor responses to icv ANG II in SHRs and 2K1C hypertensive rats treated with H_2O_2 icv or ATZ icv or iv

2K1C hypertensive rats		
Treatment	Effects	% of change
H_2O_2 icv	slight reduction of baseline MAP	around 10 mmHg
H_2O_2 icv + ANG II icv	Reduction of the pressor response	80% reduction
ATZ icv	No change of baseline MAP	-
ATZ icv + ANG II icv	Reduction of the pressor response	38% reduction
SHRs		
Treatment	Effects	% of change
H_2O_2 icv	slight reduction of baseline MAP	around 10 mmHg
H_2O_2 icv + ANG II icv	Reduction of the pressor response	55% reduction
ATZ icv	No change of baseline MAP	-
ATZ icv + ANG II icv	Reduction of the pressor response	51% reduction

H_2O_2, 5 μmol; ANG II, 50 ng; ATZ, 5 nmol.

ATZ icv reduced by 38% the pressor response to icv ANG II in 2K1C hypertensive rats (Table 2) (Lauar et al. 2020). ATZ icv produced no change in baseline MAP in 2K1C hypertensive rats. ATZ icv and ANG II icv alone or combined produced no change in HR in 2K1C hypertensive rats (Lauar et al. 2020).

These results suggest that H_2O_2 acting centrally inhibits the pressor mechanisms activated by central angiotensinergic stimulation in renal hypertensive rats and may produce a small reduction of baseline MAP acutely in these rats (Lauar et al. 2020).

Responses to Central Angiotensinergic Stimulation in Spontaneously Hypertensive Rats Treated with H_2O_2 or ATZ icv

H_2O_2 injected icv in SHRs reduced by 55% the pressor response to icv ANG II (Table 2) (Lauar et al. 2020). H_2O_2 icv also slightly reduced baseline MAP in SHRs (10 mmHg).

ATZ icv reduced by 51% the pressor response to ANG II icv in SHRs (Table 2) (Lauar et al. 2020). ATZ icv produced no change in baseline MAP in SHRs. ATZ icv and ANG II icv alone or combined produced no change in the HR in SHRs (Lauar et al. 2020).

These results suggest that H_2O_2 acting centrally inhibits the pressor mechanisms activated by central angiotensinergic stimulation, without changing baseline MAP in SHRs (Lauar et al. 2020).

Responses to Central Angiotensinergic Stimulation in Hypertensive Rats Treated with ATZ iv

ATZ (3.6 mmol/kg of body weight) iv reduced by 54% the pressor response to icv ANG II in SHRs and abolished (94% reduction) the pressor response to icv ANG II in 2K1C hypertensive rats (Table 3) (Lauar et al. 2020). ATZ iv acutely reduced baseline MAP in 2K1C hypertensive rats (around 10 mmHg), not in SHRs (Table 3) (Lauar et al. 2020). ATZ iv or ANG II icv alone or a combination produced no change in HR (Lauar et al. 2020).

These results suggest that ATZ iv also inhibits the pressor mechanisms activated by central angiotensinergic stimulation in SHRs and renal hypertensive rats. ATZ iv also reduced baseline MAP in renal hypertensive rats (Lauar et al. 2020).

Responses to Central Angiotensinergic Stimulation in Hypertensive Rats Treated with ATZ iv Combined with H_2O_2 icv

The combination of ATZ iv + H_2O_2 icv reduced by 55% the pressor response to icv ANG II in SHRs and by 78% the pressor response to icv ANG II in 2K1C hypertensive rats (Table 3) (Lauar et al. 2020).

The combination of ATZ iv + H_2O_2 icv acutely reduced (next 20 minutes) baseline MAP in SHRs and in 2K1C hypertensive rats (around 10 mmHg) (Table 3) (Lauar et al. 2020). ATZ iv or H_2O_2 icv alone or combined produced no change in HR in SHRs or 2K1C hypertensive rats, even when ANG II was injected icv (Lauar et al. 2020).

These results suggest that ATZ iv combined with H_2O_2 icv inhibits the pressor mechanisms activated by central angiotensinergic stimulation and is effective to produce an acute small reduction of baseline MAP in SHRs and renal hypertensive rats (Lauar et al. 2020).

Table 3. Baseline arterial pressure and pressor responses to icv ANG II in SHRs and 2K1C hypertensive rats treated with ATZ iv alone or combined with H_2O_2 icv

2K1C hypertensive rats		
Treatment	Effects	% of change
ATZ iv	Reduction of baseline MAP	around 10 mmHg
ATZ iv + ANG II icv	Reduction of the pressor response	94% reduction
ATZ iv + H_2O_2 icv	Reduction of MAP	around 10 mmHg
ATZ iv + H_2O_2 icv + ANG II icv	Reduction of the pressor response	78% reduction
SHRs		
Treatment	Effects	% of change
ATZ iv	No reduction of baseline MAP	-
ATZ iv + ANG II icv	Reduction of the pressor response	54% reduction
ATZ iv + H_2O_2 icv	Reduction of MAP	around 10 mmHg
ATZ iv + H_2O_2 icv + ANG II icv	Reduction of the pressor response	55% reduction

H_2O_2, 5 μmol; ANG II, 50 ng; ATZ, 3.6 mmol/kg of body weight.

Responses to Peripheral Injections of Noradrenaline or ANG II in Hypertensive Rats Treated with ATZ iv

To confirm that the anti-hypertensive effects of ATZ iv are due to its central action, and not the result of inhibition of peripheral mechanisms, the effects of acute iv ATZ on the pressor responses to noradrenaline (nor) or ANG II injected iv were tested. ATZ (3.6 mmol/kg of body weight) iv did not modify the pressor responses to noradrenaline (1 pmol/0.1 ml/rat) or ANG II (50 ng/0.1 ml/rat) injected iv in 2K1C hypertensive rats (Δ nor: 31 ± 3 mmHg and ANG II: 27 ± 3 mmHg, respectively), compared to the pressor responses to noradrenaline and ANG II iv after saline iv (Δ nor: 34 ± 3 mmHg and ANG II: 30 ± 2 mmHg, respectively) (Table 4).

ATZ iv also did not modify the pressor responses to noradrenaline or ANG II injected iv in SHRs (Δ nor: 56 ± 1 mmHg and ANG II: 47 ± 2 mmHg, respectively), compared to the pressor responses to noradrenaline and ANG II iv in these rats after saline iv (Δ nor: 51 ± 2 mmHg and ANG II: 47 ± 1 mmHg, respectively) (Table 4).

Table 4. Pressor responses to iv injections of noradrenaline and ANG II in SHRs and 2K1C hypertensive rats treated with ATZ iv

2K1C hypertensive rats		
Treatment	n	ΔMAP(mmHg)
Saline iv + noradrenaline iv	6	34 ± 3 mmHg
ATZ iv + noradrenalive iv	6	31 ± 3 mmHg
Saline iv + ANG II iv	6	30 ± 2 mmHg
ATZ iv + ANG iv	6	27 ± 3 mmHg
SHRs		
Treatment	n	ΔMAP(mmHg)
Saline iv + noradrenaline iv	5	51 ± 2 mmHg
ATZ iv + noradrenalive iv	5	56 ± 1 mmHg
Saline iv + ANG II iv	5	47 ± 1 mmHg
ATZ iv + ANG iv	5	47 ± 2 mmHg

Results are expressed as means ± SEM; n = number of rats. ANG II, 50 ng/0,1 ml/rat; ATZ, 3.6 mmol/kg of body weight; noradrenaline (1 pmol/0,1 ml/rat).

These results show that ATZ iv did not change peripheral pressor mechanisms activated by α1-adrenergic receptor or AT1 receptor activation in SHRs and renal hypertensive rats, suggesting that the effects of ATZ in the pressor response to ANG II are due to its central action.

Discussion

The results demonstrated by Lauar et al. (Lauar et al. 2010; 2019; 2020) and in the present chapter show that:

1. Icv injection of H_2O_2 or ATZ reduced the pressor response to icv ANG II in normotensive and hypertensive rats from 38 to 80% and the pressor response to icv carbachol in normotensive rats from 52 to 57%;
2. The iv injection of ATZ reduced the pressor response to icv ANG II in normotensive and hypertensive rats from 54 to 94% and the pressor response to icv carbachol in normotensive rats by 32%;
3. The combination of ATZ iv + H_2O_2 icv reduced the pressor response to icv ANG II in hypertensive rats from 55 to 78%;

4. H_2O_2 or ATZ injected icv or iv produced no change in baseline MAP in normotensive rats;
5. ATZ iv slightly reduced baseline MAP in 2K1C hypertensive rats, not in SHRs;
6. H_2O_2 injected icv alone or combined with iv ATZ produced a slight reduction (around 10 mmHg) in baseline MAP in hypertensive rats, whereas ATZ icv alone produced no change in baseline MAP in hypertensive rats;
7. ATZ iv did not change peripheral pressor mechanisms activated by α1-adrenergic receptor or AT1 receptor activation in hypertensive rats.

Together the results suggest that H_2O_2 injected icv or ATZ injected icv or iv strongly reduces pressor responses produced by central angiotensinergic or cholinergic activation in normotensive rats. H_2O_2 or ATZ injected icv and also ATZ injected iv combined with H_2O_2 icv marked reduced the pressor response to central angiotensinergic activation and also slightly reduce baseline MAP in hypertensive rats. No significant change in HR was produced by these treatments.

Effective reduction of the pressor responses was produced by central acting exogenous H_2O_2 or increased central endogenous H_2O_2 production after the blockade of catalase with ATZ, suggesting that these treatments acting centrally have effective anti-hypertensive action. Evidence of central anti-hypertensive action of these treatments is also the reduction of baseline arterial in 2K1C hypertensive rats and SHRs. Hypertension in SHRs and 2K1C hypertensive rats is characterized by increased activation of the SRA (Berenguer et al. 1991; Blanch et al. 2014; Leenen et al. 1975; Moyses et al. 1994; Oliveira-Sales et al. 2009; 2014; Tsyrlin et al. 2013; Barbosa et al. 2017). Therefore, the anti-hypertensive effects of H_2O_2 or ATZ in SHRs and 2K1C hypertensive rats are probably related to the reduction of the central action of ANG II.

The iv injection of ATZ also reduced the pressor responses and reduced baseline MAP in 2K1C hypertensive rats, again showing the anti-hypertensive action of these treatments. Also in the case of iv injection of ATZ its action is probably central. Evidence for this is the absence of the effect of iv ATZ on the pressor responses to iv ANG II or noradrenaline.

The pressor responses to icv ANG II depend on the activation of ANG II AT1 receptors located in the neurons of the circumventricular areas that release facilitatory signals to increase vasopressin secretion and sympathetic activation (Johnson, Hoffman, and Buggy 1978; Johnson 1985; Hoffman et al. 1977; Mahon et al. 1995). It is not clear which are the central areas involved in the anti-hypertensive action of H_2O_2 and ATZ. Probably they act in the same areas activated by ANG II. These areas lacking blood-brain barrier like the SFO and OVLT are close to the cerebral ventricles easily reached by H_2O_2 and ATZ injected icv. From these circumventricular sites, facilitatory signals may reach the PVN and the RVLM to increase sympathetic activity and PVN and SON to increase vasopressin secretion (Hoffman et al. 1977; Johnson, Hoffman, and Buggy 1978; Johnson 1985; Mahon et al. 1995). Although not possible to exclude other possibilities, the action in other areas of the brain circuitry involved in cardiovascular control like the PVN, NTS, and RVLM is less probably when drugs are injected icv. A study demonstrated that H_2O_2 or ATZ injected into the NTS in rats produced hypotension and bradycardia that were reduced by the injection of the glutamate antagonist kynurenate, suggesting that H_2O_2 into the NTS induces hypotension and bradycardia probably due to activation of glutamatergic mechanisms (Cardoso et al. 2009). Unlike injections of H_2O_2 or ATZ into the NTS, the injection of H_2O_2 into the 4th cerebral ventricle produced transient pressor responses and long lasting bradycardia that were suggested to depend on simultaneous activation of sympathetic and parasympathetic activity (Máximo Cardoso et al. 2006). Contrary to the results of the studies of Cardoso et al., no significant effect of H_2O_2 or ATZ injected iv or into the lateral ventricle was observed on baseline MAP or HR in normotensive rats, which suggests that H_2O_2 acting in different areas of the brain may produce different effects in the mechanisms of cardiovascular control.

Another study demonstrated that injections of H_2O_2 into the medial septal area (MSA) reduced water intake, antidiuresis, natriuresis, pressor responses, and the number of vasopressinergic neurons expressing c-Fos in the PVN and oxytocinergic neurons expressing c-Fos in the PVN and in the SON to carbachol injected into the MSA (Melo et al. 2015). The previous treatment with the K^+ ATP channel blocker glibenclamide in the MSA partially reversed the inhibitory responses produced by H_2O_2, suggesting that H_2O_2 inhibits responses produced by cholinergic activation of the MSA acting through K^+ ATP channels in this area (Melo et al. 2015). A study also showed that ATZ injected into the MSA partially reduced the antidiuretic effect of carbachol injected into the same area (Sá et al. 2019). In addition, the combination of ATZ and H_2O_2 into the MSA abolished the antidiuresis and reduced water intake induced by carbachol injected into the same area (Sá et al. 2019). The studies showing hypotension and bradycardia after injections of H_2O_2 or ATZ into the NTS and inhibition of responses produced by carbachol injected into the MSA reinforce the suggestion that exogenous or endogenous H_2O_2 modulates central mechanisms that control cardiovascular and fluid electrolytic response. In addition, hyperpolarization and decreased action potential discharge in neurons of the PVN with the treatment with H_2O_2 were described, responses that were blocked by K^+ channel blockers glibenclamide reinforcing the importance of K^+ ATP channels for the central effects of H_2O_2, particularly in areas involved in cardiovascular regulation like the MSA and PVN (Dantzler et al. 2019).

Acting centrally, H_2O_2 may activate different mechanisms to inhibit ANG II and carbachol action. H_2O_2 may affect neuronal excitability through different mechanisms like changes in neurotransmitter release or ion channel activation. H_2O_2 can block glutamate uptake by glial cells, which may result in an increase in extracellular glutamate levels and enhanced neuronal excitability (Sorg et al. 1997; Volterra et al. 1994). In opposite direction, H_2O_2 may also reduce neuronal excitability as a consequence of the reduction of glutamate or increase in GABA release or by the activation of K_{ATP} channels (Zoccarato, Valente, and Alexandre 1995; Zoccarato et al. 1999; Sah et al. 2002; Takahashi, Mikami, and Yang 2007; Bao, Avshalumov, and Rice 2005; Avshalumov et al. 2005). In this chapter, it is reported that increased levels of H_2O_2 centrally reduce the facilitatory signals produced by central ANG II to induce pressor responses, probably decreasing the neuronal activity in the SFO and/or OVLT to reduce the pressor response to central ANG II.

The release of $O_2^{\bullet-}$ by the activation of central AT1 receptors is part of the signaling mechanisms activated by ANG II to produce pressor and dipsogenic responses (Zimmerman et al. 2002; Zimmerman, Lazartigues, et al. 2004; Zimmerman and Davisson 2004). This suggestion is based on studies that demonstrated that central SOD overexpression reducing $O_2^{\bullet-}$ levels centrally also reduces the pressor response to central ANG II (Zimmerman et al. 2002; Zimmerman, Lazartigues, et al. 2004; Zimmerman, Dunlay, et al. 2004; Zimmerman and Davisson 2004). However, the dismutation of $O_2^{\bullet-}$ by SOD releases H_2O_2. In spite that changes in SOD activity do not necessarily affect H_2O_2 levels (Teixeira, Schumacher, and Meneghini 1998; Gardner, Salvador, and Moradas-Ferreira 2002; Chan et al. 2006; Kowald, Lehrach, and Klipp 2006), SOD overexpression may result in high levels of H_2O_2 release as a consequence of the expressive dismutation of $O_2^{\bullet-}$, increasing H_2O_2 to levels enough to reduce the pressor responses to central ANG II as reported in the present chapter. On the other hand, the degradation of H_2O_2 by catalase activity is a limitation for H_2O_2 produced by dismutation of $O_2^{\bullet-}$ in physiological conditions to produce strong inhibition of ANG II actions, which means that in physiological conditions H_2O_2 may only transiently inhibit ANG II. The treatment with ATZ reduces the activity of catalase and increases the levels of endogenously produced H_2O_2. And the results of catalase blockade with ATZ are a reduction of ANG II effects similar to the treatment with high levels of exogenous H_2O_2 as reported in the present chapter. The possible involvement of H_2O_2 in the inhibition of ANG II-induced pressor and dipsogenic responses by SOD overexpression is a question to be tested in future studies. The results showing that ATZ treatment also did not modify the pressor responses produced by iv injections of noradrenaline or ANG II in 2K1C hypertensive rats or SHRs also suggest that the anti-hypertensive effects of ATZ are not due to a vascular action causing impairment of pressor mechanisms activated by these pressor agents. Instead, the effects of ATZ are suggested to be due to its action blocking catalase centrally.

Important to note that, while H_2O_2 or ATZ injected into the lateral ventricle does not change baseline MAP in normotensive rats, in hypertensive rats, the same injection into the lateral ventricle slightly reduced baseline MAP, without changes in HR demonstrating that ANG II plays important role in the models of hypertension tested (Berenguer et al. 1991; Blanch et al. 2014; Leenen et al. 1975; Moyses et al. 1994; Oliveira-Sales et al. 2009; 2014; Tsyrlin et al. 2013; Veerasingham and Raizada 2003; Tanaka et al. 1995). Therefore, the inhibition of ANG II action by the treatment with H_2O_2 or ATZ in hypertensive rats is probably the mechanism involved in the reduction of baseline MAP.

Conclusion

The study examined the role of central exogenous injections of H_2O_2 and central or peripheral injections of catalase inhibitor (ATZ) in increasing endogenous levels of H_2O_2 on the pressor responses induced by icv ANG II or carbachol. The results suggest that exogenous or endogenous H_2O_2 may inhibit central pressor mechanisms (sympathoactivation and/or vasopressin release) activated by central ANG II or cholinergic stimulation. Moreover, exogenous or endogenous H_2O_2 acting centrally produces anti-hypertensive effects impairing central pressor mechanisms activated by ANG II in SHRs or 2K1C hypertensive rats. Therefore, increases in H_2O_2 due to inhibition of catalase enzyme might produce beneficial effects in sort of hypertension which has angiotensinergic and/or cholinergic mechanisms involved. Figure 1 is a schematic model showing how the blockade of catalase increasing H_2O_2 levels affects the balance between $O_2^{\bullet-}$ and H_2O_2 to modulate central angiotensinergic and cholinergic mechanisms that activate AVP release and sympathetic activity to produce pressor responses and hypertension. According to previous studies (Zimmerman et al. 2002; Zimmerman and Davisson 2004; Zimmerman, Dunlay, et al. 2004), $O_2^{\bullet-}$ is part of the signaling mechanisms activated by ANG II centrally (positive modulation), whereas increased levels of H_2O_2 that result from catalase blockade with ATZ or exogenous H_2O_2 administration produce opposite effects (negative modulation) of central angiotensinergic and cholinergic activity.

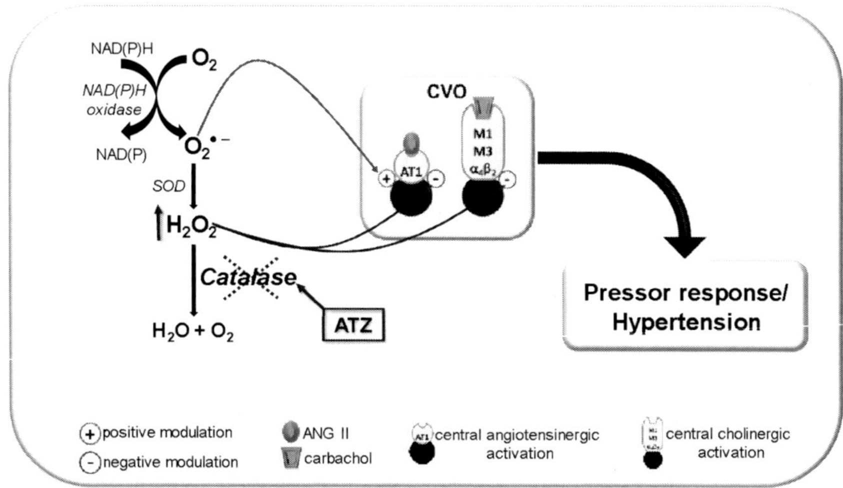

Figure 1. Schematic model showing how changes in catalase activity/H_2O_2 levels may affect the activity of central angiotensinergic and cholinergic mechanisms.

References

Adler, V, Z Yin, KD Tew, and Z Ronai. 1999. "Role of Redox Potential and Reactive Oxygen Species in Stress Signaling." *Oncogene* 18 (45): 6104–11. https://doi.org/10.1038/sj.onc.1203128.

Aizenman, E, S A Lipton, and R H Loring. 1989. "Selective Modulation of NMDA Responses by Reduction and Oxidation." *Neuron* 2 (3): 1257–63. http://www.ncbi.nlm.nih.gov/pubmed/2696504.

Álvarez, Ezequiel, Beatriz Rodiño-Janeiro, María Isabel Paradela-Dobarro, Sergio Castiñeiras-Landeira, José R Raposeiras-Roubín, and Ezequiel Gonzalez-Juanatey.

2013. "Current Status of NADPH Oxidase Research in Cardiovascular Pharmacology." *Vascular Health and Risk Management* 9 (July): 401. https://doi.org/10.2147/VHRM.S33053.

Aragon, CM, F Rogan, and Z Amit. 1991. "Dose- and Time-Dependent Effect of an Acute 3-Amino-1,2,4-Triazole Injection on Rat Brain Catalase Activity." *Biochemical Pharmacology* 42 (3): 699–702. http://www.ncbi.nlm.nih.gov/pubmed/1859472.

Aronsson, M, K Almasan, K Fuxe, A Cintra, A Härfstrand, JA Gustafsson, and D Ganten. 1988. "Evidence for the Existence of Angiotensinogen MRNA in Magnocellular Paraventricular Hypothalamic Neurons." *Acta Physiologica Scandinavica* 132 (4): 585–86. https://doi.org/10.1111/j.1748-1716.1988.tb08370.x.

Avshalumov, Marat V, Billy T Chen, Tibor Koós, James M Tepper, and Margaret E Rice. 2005. "Endogenous Hydrogen Peroxide Regulates the Excitability of Midbrain Dopamine Neurons via ATP-Sensitive Potassium Channels." *The Journal of Neuroscience : The Official Journal of the Society for Neuroscience* 25 (17): 4222–31. https://doi.org/10.1523/JNEUROSCI.4701-04.2005.

Bao, Li, Marat V Avshalumov, Jyoti C Patel, Christian R Lee, Evan W Miller, Christopher J Chang, and Margaret E Rice. 2009. "Mitochondria Are the Source of Hydrogen Peroxide for Dynamic Brain-Cell Signaling." *The Journal of Neuroscience : The Official Journal of the Society for Neuroscience* 29 (28): 9002–10. https://doi.org/10.1523/JNEUROSCI.1706-09.2009.

Bao, Li, Marat V Avshalumov, and Margaret E Rice. 2005. "Partial Mitochondrial Inhibition Causes Striatal Dopamine Release Suppression and Medium Spiny Neuron Depolarization via H_2O_2 Elevation, Not ATP Depletion." *The Journal of Neuroscience : The Official Journal of the Society for Neuroscience* 25 (43): 10029–40. https://doi.org/10.1523/JNEUROSCI.2652-05.2005.

Barbosa, Rafaela Moreira, Guilherme F. Speretta, Daniel Penteado Martins Dias, Prashant Jay Ruchaya, Hongwei Li, José Vanderlei Menani, Colin Sumners, Eduardo Colombari, and Débora S. A. Colombari. 2017. "Increased Expression of Macrophage Migration Inhibitory Factor in the Nucleus of the Solitary Tract Attenuates Renovascular Hypertension in Rats." *American Journal of Hypertension*, January. https://doi.org/10.1093/ajh/hpx001.

Berenguer, LM, J Garcia-Estañ, M Ubeda, AJ Ortiz, and T Quesada. 1991. "Role of Renin-Angiotensin System in the Impairment of Baroreflex Control of Heart Rate in Renal Hypertension." *Journal of Hypertension* 9 (12): 1127–33. http://www.ncbi.nlm.nih.gov/pubmed/1663968.

Berry, C, CA Hamilton, MJ Brosnan, FG Magill, GA Berg, JJ McMurray, and AF Dominiczak. 2000. "Investigation into the Sources of Superoxide in Human Blood Vessels: Angiotensin II Increases Superoxide Production in Human Internal Mammary Arteries." *Circulation* 101 (18): 2206–12. http://www.ncbi.nlm.nih.gov/pubmed/10801763.

Blair-West, JR, KD Carey, DA Denton, RS Weisinger, and RE Shade. 1998. "Evidence That Brain Angiotensin II Is Involved in Both Thirst and Sodium Appetite in Baboons." *The American Journal of Physiology* 275 (5 Pt 2): R1639-46. http://www.ncbi.nlm.nih.gov/pubmed/9791085.

Blanch, Graziela Torres, André Henrique Freiria-Oliveira, Guilherme Fleury Fina Speretta, Eduardo J Carrera, Hongwei Li, Robert C Speth, Eduardo Colombari, Colin Sumners, and Débora S A Colombari. 2014. "Increased Expression of Angiotensin II Type 2 Receptors in the Solitary-Vagal Complex Blunts Renovascular Hypertension." *Hypertension* 64 (4): 777–83. https://doi.org/10.1161/HYPERTENSIONAHA.114.03188.

Buggy, J, GD Fink, AK Johnson, and MJ Brody. 1977. "Prevention of the Development of Renal Hypertension by Anteroventral Third Ventricular Tissue Lesions." *Circulation Research* 40 (5 Suppl 1): I110-7. http://www.ncbi.nlm.nih.gov/pubmed/858173.

Cardoso, LM, DSd. A. Colombari, JV. Menani, GM. Toney, DA. Chianca, and E Colombari. 2009. "Cardiovascular Responses to Hydrogen Peroxide into the Nucleus Tractus Solitarius." *AJP: Regulatory, Integrative and Comparative Physiology* 297 (2): R462–69. https://doi.org/10.1152/ajpregu.90796.2008.

Chan, Samuel HH, Ming-Hon Tai, Chia-Yen Li, and Julie YH Chan. 2006. "Reduction in Molecular Synthesis or Enzyme Activity of Superoxide Dismutases and Catalase Contributes to Oxidative Stress and Neurogenic Hypertension in Spontaneously Hypertensive Rats." *Free Radical Biology & Medicine* 40 (11): 2028–39. https://doi.org/10.1016/j.freeradbiomed.2006.01.032.

Chen, BT, MV Avshalumov, and ME Rice. 2001. "H(2)O(2) Is a Novel, Endogenous Modulator of Synaptic Dopamine Release." *Journal of Neurophysiology* 85 (6): 2468–76. http://www.ncbi.nlm.nih.gov/pubmed/11387393.

Cohen, G. 1994. "Enzymatic/Nonenzymatic Sources of Oxyradicals and Regulation of Antioxidant Defenses." *Annals of the New York Academy of Sciences* 738 (November): 8–14. http://www.ncbi.nlm.nih.gov/pubmed/7832459.

Correa, FM, LM Plunkett, and JM Saavedra. 1986. "Quantitative Distribution of Angiotensin-Converting Enzyme (Kininase II) in Discrete Areas of the Rat Brain by Autoradiography with Computerized Microdensitometry." *Brain Research* 375 (2): 259–66. http://www.ncbi.nlm.nih.gov/pubmed/3015330.

Cowley, AW, and AC Guyton. 1972. "Quantification of Intermediate Steps in the Renin-Angiotensin-Vasoconstrictor Feedback Loop in the Dog." *Circulation Research* 30 (5): 557–66. http://www.ncbi.nlm.nih.gov/pubmed/4337381.

Dantzler, Heather A, Michael P Matott, Diana Martinez, and David D Kline. 2019. "Hydrogen Peroxide Inhibits Neurons in the Paraventricular Nucleus of the Hypothalamus via Potassium Channel Activation." *American Journal of Physiology-Regulatory, Integrative and Comparative Physiology* 317 (1): R121–33. https://doi.org/10.1152/ajpregu.00054.2019.

Deschepper, CF, J Bouhnik, and WF Ganong. 1986. "Colocalization of Angiotensinogen and Glial Fibrillary Acidic Protein in Astrocytes in Rat Brain." *Brain Research* 374 (1): 195–98. http://www.ncbi.nlm.nih.gov/pubmed/3521789.

Dzau, VJ, J Ingelfinger, RE Pratt, and KE Ellison. 1986. "Identification of Renin and Angiotensinogen Messenger RNA Sequences in Mouse and Rat Brains." *Hypertension (Dallas, Tex. : 1979)* 8 (6): 544–48. http://www.ncbi.nlm.nih.gov/pubmed/3519452.

Fitzsimons, JT. 1998. "Angiotensin, Thirst, and Sodium Appetite." *Physiological Reviews* 78 (3): 583–686. http://www.ncbi.nlm.nih.gov/pubmed/9674690.

Ganten, D, JL Minnich, P Granger, K Hayduk, HM Brecht, A Barbeau, R Boucher, and J Genest. 1971. "Angiotensin-Forming Enzyme in Brain Tissue." *Science (New York, N.Y.)* 173 (3991): 64–65. http://www.ncbi.nlm.nih.gov/pubmed/4325865.

Gardner, Rui, Armindo Salvador, and Pedro Moradas-Ferreira. 2002. "Why Does SOD Overexpression Sometimes Enhance, Sometimes Decrease, Hydrogen Peroxide Production? A Minimalist Explanation." *Free Radical Biology & Medicine* 32 (12): 1351–57. http://www.ncbi.nlm.nih.gov/pubmed/12057773.

Griendling, KK, CA Minieri, JD Ollerenshaw, and RW Alexander. 1994. "Angiotensin II Stimulates NADH and NADPH Oxidase Activity in Cultured Vascular Smooth Muscle Cells." *Circulation Research* 74 (6): 1141–48. http://www.ncbi.nlm.nih.gov/pubmed/8187280.

Grobe, JL., D Xu, and CD. Sigmund. 2008. "An Intracellular Renin-Angiotensin System in Neurons: Fact, Hypothesis, or Fantasy." *Physiology* 23 (4): 187–93. https://doi.org/10.1152/physiol.00002.2008.

Hackenthal, E, M Paul, D Ganten, and R Taugner. 1990. "Morphology, Physiology, and Molecular Biology of Renin Secretion." *Physiological Reviews* 70 (4): 1067–1116. http://www.ncbi.nlm.nih.gov/pubmed/2217555.

Heim, WG, D Appleman, and HT Pyfrom. 1955. "Production of Catalase Changes in Animals with 3-Amino-1, 2, 4-Triazole." *Science (New York, N.Y.)* 122 (3172): 693–94. http://www.ncbi.nlm.nih.gov/pubmed/13255910.

Hodge, RL, RD Lowe, and JR Vane. 1966. "The Effects of Alteration of Blood-Volume on the Concentration of Circulating Angiotensin in Anaesthetized Dogs." *The Journal of Physiology* 185 (3): 613–26. http://www.ncbi.nlm.nih.gov/pubmed/4288206.

Hoffman, WE, MI Philips, PG Schmid, J Falcon, and JF Weet. 1977. "Antidiuretic Hormone Release and the Pressor Response to Central Angiotensin II and Cholinergic Stimulation." *Neuropharmacology* 16 (7–8): 463–72. http://www.ncbi.nlm.nih.gov/pubmed/917252.

Irani, K. 2000. "Oxidant Signaling in Vascular Cell Growth, Death, and Survival : A Review of the Roles of Reactive Oxygen Species in Smooth Muscle and Endothelial Cell Mitogenic and Apoptotic Signaling." *Circulation Research* 87 (3): 179–83. http://www.ncbi.nlm.nih.gov/pubmed/10926866.

Jiang, B, JH Liu, YM Bao, and LJ An. 2003. "Hydrogen Peroxide-Induced Apoptosis in Pc12 Cells and the Protective Effect of Puerarin." *Cell Biology International* 27 (12): 1025–31. http://www.ncbi.nlm.nih.gov/pubmed/14642535.

Johnson, AK. 1985. "The Periventricular Anteroventral Third Ventricle (AV3V): Its Relationship with the Subfornical Organ and Neural Systems Involved in Maintaining Body Fluid Homeostasis." *Brain Research Bulletin* 15 (6): 595–601. http://www.ncbi.nlm.nih.gov/pubmed/3910170.

Johnson, AK, WE Hoffman, and J Buggy. 1978. "Attenuated Pressor Responses to Intracranially Injected Stimuli and Altered Antidiuretic Activity Following Preoptic-Hypothalamic Periventricular Ablation." *Brain Research* 157 (1): 161–66. http://www.ncbi.nlm.nih.gov/pubmed/698843.

Kloet, Annette D de, Meng Liu, Vermalí Rodríguez, Eric G Krause, and Colin Sumners. 2015. "Role of Neurons and Glia in the CNS Actions of the Renin-Angiotensin System in Cardiovascular Control." *American Journal of Physiology - Regulatory, Integrative and Comparative Physiology* 309 (5): R444–58. https://doi.org/10.1152/ajpregu.00078.2015.

Kowald, Axel, Hans Lehrach, and Edda Klipp. 2006. "Alternative Pathways as Mechanism for the Negative Effects Associated with Overexpression of Superoxide Dismutase." *Journal of Theoretical Biology* 238 (4): 828–40. https://doi.org/10.1016/j.jtbi.2005.06.034.

Lassègue, Bernard, and Roza E. Clempus. 2003. "Vascular NAD(P)H Oxidases: Specific Features, Expression, and Regulation." *American Journal of Physiology - Regulatory, Integrative and Comparative Physiology* 285 (2): R277–97. https://doi.org/10.1152/ajpregu.00758.2002.

Lauar, Mariana R., DSA Colombari, PM De Paula, E Colombari, LM Cardoso, LA De Luca, and JV Menani. 2010. "Inhibition of Central Angiotensin II-Induced Pressor Responses by Hydrogen Peroxide." *Neuroscience* 171 (2): 524–30. https://doi.org/10.1016/j.neuroscience.2010.08.048.

Lauar, Mariana R., Débora SA Colombari, Eduardo Colombari, Patrícia M De Paula, Laurival A De Luca, and José V Menani. 2019. "Catalase Blockade Reduces the Pressor Response to Central Cholinergic Activation." *Brain Research Bulletin* 153 (November): 266–72. https://doi.org/10.1016/J.BRAINRESBULL.2019.09.008.

Lauar, Mariana R, Graziela T Blanch, Débora SA Colombari, Eduardo Colombari, Patrícia M De Paula, Laurival A De Luca, and José V Menani. 2020. "Anti-Hypertensive Effect of Hydrogen Peroxide Acting Centrally." *Hypertension Research : Official Journal of the Japanese Society of Hypertension*, May. https://doi.org/10.1038/s41440-020-0474-5.

Leenen, FH, JW Scheeren, D Omylanowski, JD Elema, B Van der Wal, and W De Jong. 1975. "Changes in the Renin-Angiotensin-Aldosterone System and in Sodium and Potassium Balance during Development of Renal Hypertension in Rats." *Clinical Science and Molecular Medicine* 48 (1): 17–26. http://www.ncbi.nlm.nih.gov/pubmed/1167491.

Liang, Chang-Seng, and Haralambos Gavras. 1978. "Renin-Angiotensin System Inhibition in Conscious Dogs during Acute Hypoxemia." *Journal of Clinical Investigation* 62 (5): 961–70. https://doi.org/10.1172/JCI109225.

Mahon, JM, M Allen, J Herbert, and JT Fitzsimons. 1995. "The Association of Thirst, Sodium Appetite and Vasopressin Release with c-Fos Expression in the Forebrain of the Rat after Intracerebroventricular Injection of Angiotensin II, Angiotensin-(1-7) or Carbachol." *Neuroscience* 69 (1): 199–208. http://www.ncbi.nlm.nih.gov/pubmed/8637618.

Maker, HS, C Weiss, DJ Silides, and G Cohen. 1981. "Coupling of Dopamine Oxidation (Monoamine Oxidase Activity) to Glutathione Oxidation via the Generation of Hydrogen Peroxide in Rat Brain Homogenates." *Journal of Neurochemistry* 36 (2): 589–93. http://www.ncbi.nlm.nih.gov/pubmed/7463078.

Matsumura, K, DB Averill, and CM Ferrario. 1998. "Angiotensin II Acts at AT1 Receptors in the Nucleus of the Solitary Tract to Attenuate the Baroreceptor Reflex." *The American Journal of Physiology* 275 (5 Pt 2): R1611-9. http://www.ncbi.nlm.nih.gov/pubmed/9791081.

Máximo Cardoso, Leonardo, Débora Simões de Almeida Colombari, José Vanderlei Menani, Deoclécio Alves Chianca, and Eduardo Colombari. 2006. "Cardiovascular Responses Produced by Central Injection of Hydrogen Peroxide in Conscious Rats." *Brain Research Bulletin* 71 (1–3): 37–44. https://doi.org/10.1016/j.brainresbull.2006.07.013.

McCord, JM, and I Fridovich. 1969. "The Utility of Superoxide Dismutase in Studying Free Radical Reactions. I. Radicals Generated by the Interaction of Sulfite, Dimethyl Sulfoxide, and Oxygen." *The Journal of Biological Chemistry* 244 (22): 6056–63. http://www.ncbi.nlm.nih.gov/pubmed/4981789.

Melo, MR, JV Menani, E Colombari, and DSA Colombari. 2015. "Hydrogen Peroxide Attenuates the Dipsogenic, Renal and Pressor Responses Induced by Cholinergic Activation of the Medial Septal Area." *Neuroscience* 284 (January): 611–21. https://doi.org/10.1016/j.neuroscience.2014.10.024.

Michelini, LC, and LG Bonagamba. 1990. "Angiotensin II as a Modulator of Baroreceptor Reflexes in the Brainstem of Conscious Rats." *Hypertension (Dallas, Tex. : 1979)* 15 (2 Suppl): I45-50. http://www.ncbi.nlm.nih.gov/pubmed/2298476.

Moyses, MR, AM Cabral, D Marçal, and EC Vasquez. 1994. "Sigmoidal Curve-Fitting of Baroreceptor Sensitivity in Renovascular 2K1C Hypertensive Rats." *Brazilian Journal of Medical and Biological Research = Revista Brasileira de Pesquisas Médicas e Biológicas / Sociedade Brasileira de Biofísica ... [et Al.]* 27 (6): 1419–24. http://www.ncbi.nlm.nih.gov/pubmed/7894357.

Navar, L Gabriel, Lixian Zou, Annette Von Thun, Chi Tarng Wang, John D Imig, and Kenneth D. Mitchell. 1998. "Unraveling the Mystery of Goldblatt Hypertension." *News in Physiological Sciences : An International Journal of Physiology Produced Jointly by the International Union of Physiological Sciences and the American Physiological Society* 13 (August): 170–76. http://www.ncbi.nlm.nih.gov/pubmed/11390784.

Nicholls, P. 1962. "The Reaction between Aminotriazole and Catalase." *Biochimica et Biophysica Acta* 59 (May): 414–20. http://www.ncbi.nlm.nih.gov/pubmed/14479447.

Oliveira-Sales, Elizabeth B, Marie Ann Toward, Ruy R Campos, and Julian FR Paton. 2014. "Revealing the Role of the Autonomic Nervous System in the Development and Maintenance of Goldblatt Hypertension in Rats." *Autonomic Neuroscience* 183 (July): 23–29. https://doi.org/10.1016/j.autneu.2014.02.001.

Oliveira-Sales, Elizabeth B, Erika E Nishi, Bruno A Carillo, Mirian A Boim, Miriam S Dolnikoff, Cássia T Bergamaschi, and Ruy R Campos. 2009. "Oxidative Stress in the Sympathetic Premotor Neurons Contributes to Sympathetic Activation in Renovascular Hypertension." *American Journal of Hypertension* 22 (5): 484–92. https://doi.org/10.1038/ajh.2009.17.

Palkovits, M, and L Záborszky. 1977. "Neuroanatomy of Central Cardiovascular Control. Nucleus Tractus Solitarii: Afferent and Efferent Neuronal Connections in Relation to the Baroreceptor Reflex Arc." *Progress in Brain Research* 47: 9–34. https://doi.org/10.1016/S0079-6123(08)62709-0.

Paton, JF, and S Kasparov. 1999. "Differential Effects of Angiotensin II on Cardiorespiratory Reflexes Mediated by Nucleus Tractus Solitarii - a Microinjection

Study in the Rat." *The Journal of Physiology* 521 Pt 1 (November): 213–25. http://www.ncbi.nlm.nih.gov/pubmed/10562346.

Peterson, Jeffrey R, Melissa A Burmeister, Xin Tian, Yi Zhou, Mallikarjuna R Guruju, John A Stupinski, Ram V Sharma, and Robin L Davisson. 2009. "Genetic Silencing of Nox2 and Nox4 Reveals Differential Roles of These NADPH Oxidase Homologues in the Vasopressor and Dipsogenic Effects of Brain Angiotensin II." *Hypertension (Dallas, Tex. : 1979)* 54 (5): 1106–14. https://doi.org/10.1161/HYPERTENSION AHA.109.140087.

Phillips, MI, and C Sumners. 1998. "Angiotensin II in Central Nervous System Physiology." *Regulatory Peptides* 78 (1–3): 1–11. http://www.ncbi.nlm.nih.gov/pubmed/9879741.

Rhee, Sue Goo, Tong-Shin Chang, Yun Soo Bae, Seung-Rock Lee, and Sang Won Kang. 2003. "Cellular Regulation by Hydrogen Peroxide." *Journal of the American Society of Nephrology : JASN* 14 (8 Suppl 3): S211-5. http://www.ncbi.nlm.nih.gov/pubmed/12874433.

Rice, Margaret E. 2011. "H2O2: A Dynamic Neuromodulator." *The Neuroscientist : A Review Journal Bringing Neurobiology, Neurology and Psychiatry* 17 (4): 389–406. https://doi.org/10.1177/1073858411404531.

Sá, Jéssica Matheus, Milena Cassolatti Barros, Mariana Rosso Melo, Eduardo Colombari, José Vanderlei Menani, and Débora Simões Almeida Colombari. 2019. "Endogenous Hydrogen Peroxide Affects Antidiuresis to Cholinergic Activation in the Medial Septal Area." *Neuroscience Letters* 694 (February): 51–56. https://doi.org/10.1016/J.NEULET.2018.11.023.

Saavedra, JM, J Fernandez-Pardal, and C Chevillard. 1982. "Angiotensin-Converting Enzyme in Discrete Areas of the Rat Forebrain and Pituitary Gland." *Brain Research* 245 (2): 317–25. http://www.ncbi.nlm.nih.gov/pubmed/6289966.

Sah, Renu, Francesca Galeffi, Rebecca Ahrens, Gwendolyn Jordan, and Rochelle D Schwartz-Bloom. 2002. "Modulation of the GABA(A)-Gated Chloride Channel by Reactive Oxygen Species." *Journal of Neurochemistry* 80 (3): 383–91. http://www.ncbi.nlm.nih.gov/pubmed/11905987.

Santos RAS & Sampaio WO. 2002. "Revista Da Sociedade Brasileira de Hipertensão," 47–51.

Sorg, O, TF Horn, N Yu, DL Gruol, and FE Bloom. 1997. "Inhibition of Astrocyte Glutamate Uptake by Reactive Oxygen Species: Role of Antioxidant Enzymes." *Molecular Medicine (Cambridge, Mass.)* 3 (7): 431–40. http://www.pubmedcentral.nih.gov/articlerender.fcgi?artid=2230215&tool=pmcentrez&rendertype=abstract.

Stornetta, RL, CL Hawelu-Johnson, PG Guyenet, and KR Lynch. 1988. "Astrocytes Synthesize Angiotensinogen in Brain." *Science (New York, N.Y.)* 242 (4884): 1444–46. http://www.ncbi.nlm.nih.gov/pubmed/3201232.

Sun, Chengwen, Kathleen W Sellers, Colin Sumners, and Mohan K Raizada. 2005. "NAD(P)H Oxidase Inhibition Attenuates Neuronal Chronotropic Actions of Angiotensin II." *Circulation Research* 96 (6): 659–66. https://doi.org/10.1161/01.RES.0000161257.02571.4b.

Takahashi, Ayako, Maya Mikami, and Jay Yang. 2007. "Hydrogen Peroxide Increases GABAergic MIPSC through Presynaptic Release of Calcium from IP3 Receptor-

Sensitive Stores in Spinal Cord Substantia Gelatinosa Neurons." *The European Journal of Neuroscience* 25 (3): 705–16. https://doi.org/10.1111/j.1460-9568.2007.05323.x.

Tanaka, Junichi, Yutaka Yamamuro, Hideo Saito, Michihiko Matsuda, and Masahiko Nomura. 1995. "Differences in Electrophysiological Properties of Angiotensinergic Pathways from the Subfornical Organ to the Median Preoptic Nucleus between Normotensive Wistar-Kyoto and Spontaneously Hypertensive Rats." *Experimental Neurology* 134 (2): 192–98. https://doi.org/10.1006/EXNR.1995.1048.

Teixeira, HD, RI Schumacher, and R Meneghini. 1998. "Lower Intracellular Hydrogen Peroxide Levels in Cells Overexpressing CuZn-Superoxide Dismutase." *Proceedings of the National Academy of Sciences of the United States of America* 95 (14): 7872–75. http://www.pubmedcentral.nih.gov/articlerender.fcgi?artid=20896&tool=pmcentrez&rendertype=abstract.

Torvik, A. 1956. "Afferent Connections to the Sensory Trigeminal Nuclei, the Nucleus of the Solitary Tract and Adjacent Structures; an Experimental Study in the Rat." *The Journal of Comparative Neurology* 106 (1): 51–141. http://www.ncbi.nlm.nih.gov/pubmed/13398491.

Tsyrlin, Vitaly A, Michael M Galagudza, Nataly V Kuzmenko, Michael G Pliss, Nataly S Rubanova, and Yury I Shcherbin. 2013. "Arterial Baroreceptor Reflex Counteracts Long-Term Blood Pressure Increase in the Rat Model of Renovascular Hypertension." *PloS One* 8 (6): e64788. https://doi.org/10.1371/journal.pone.0064788.

Vaziri, ND, XQ Wang, Z Ni Ni, S Kivlighn, and S Shahinfar. 2002. "Effects of Aging and AT-1 Receptor Blockade on NO Synthase Expression and Renal Function in SHR." *Biochimica et Biophysica Acta* 1592 (2): 153–61. http://www.ncbi.nlm.nih.gov/pubmed/12379478.

Veerasingham, Shereeni J, and Mohan K Raizada. 2003. "Brain Renin-Angiotensin System Dysfunction in Hypertension: Recent Advances and Perspectives." *British Journal of Pharmacology* 139 (2): 191–202. https://doi.org/10.1038/sj.bjp.0705262.

Volterra, A, D Trotti, C Tromba, S Floridi, and G Racagni. 1994. "Glutamate Uptake Inhibition by Oxygen Free Radicals in Rat Cortical Astrocytes." *The Journal of Neuroscience : The Official Journal of the Society for Neuroscience* 14 (5 Pt 1): 2924–32. http://www.ncbi.nlm.nih.gov/pubmed/7910203.

Wehage, Edith, Jörg Eisfeld, Inka Heiner, Eberhard Jüngling, Christof Zitt, and Andreas Lückhoff. 2002. "Activation of the Cation Channel Long Transient Receptor Potential Channel 2 (LTRPC2) by Hydrogen Peroxide. A Splice Variant Reveals a Mode of Activation Independent of ADP-Ribose." *The Journal of Biological Chemistry* 277 (26): 23150–56. https://doi.org/10.1074/jbc.M112096200.

Weinberg, ED. 1990. "Cellular Iron Metabolism in Health and Diseased." *Drug Metabolism Reviews* 22 (5): 531–79. https://doi.org/10.3109/03602539008991450.

Wright, John W, and Joseph W Harding. 2013. "The Brain Renin-Angiotensin System: A Diversity of Functions and Implications for CNS Diseases." *Pflugers Archiv : European Journal of Physiology* 465 (1): 133–51. https://doi.org/10.1007/s00424-012-1102-2.

Zhu, Guo-Qing, Lie Gao, Kuashik P Patel, Irving H Zucker, and Wei Wang. 2004. "ANG II in the Paraventricular Nucleus Potentiates the Cardiac Sympathetic Afferent Reflex in Rats with Heart Failure." *Journal of Applied Physiology (Bethesda, Md. : 1985)* 97 (5): 1746–54. https://doi.org/10.1152/japplphysiol.00573.2004.

Zhu, M, CH Gelband, P Posner, and C Sumners. 1999. "Angiotensin II Decreases Neuronal Delayed Rectifier Potassium Current: Role of Calcium/Calmodulin-Dependent Protein Kinase II." *Journal of Neurophysiology* 82 (3): 1560–68. http://www.ncbi.nlm.nih.gov/pubmed/10482769.

Zimmerman, Matthew C, and Robin L Davisson. 2004. "Redox Signaling in Central Neural Regulation of Cardiovascular Function." *Progress in Biophysics and Molecular Biology* 84 (2–3): 125–49. https://doi.org/10.1016/j.pbiomolbio.2003.11.009.

Zimmerman, Matthew C, Ryan P Dunlay, Eric Lazartigues, Yulong Zhang, Ram V Sharma, John F Engelhardt, and Robin L Davisson. 2004. "Requirement for Rac1-Dependent NADPH Oxidase in the Cardiovascular and Dipsogenic Actions of Angiotensin II in the Brain." *Circulation Research* 95 (5): 532–39. https://doi.org/10.1161/01.RES.0000139957.22530.b9.

Zimmerman, Matthew C, Eric Lazartigues, Julie A Lang, Puspha Sinnayah, Iman M Ahmad, Douglas R Spitz, and Robin L Davisson. 2002. "Superoxide Mediates the Actions of Angiotensin II in the Central Nervous System." *Circulation Research* 91 (11): 1038–45. http://www.ncbi.nlm.nih.gov/pubmed/12456490.

Zimmerman, Matthew C, Eric Lazartigues, Ram V Sharma, and Robin L Davisson. 2004. "Hypertension Caused by Angiotensin II Infusion Involves Increased Superoxide Production in the Central Nervous System." *Circulation Research* 95 (2): 210–16. https://doi.org/10.1161/01.RES.0000135483.12297.e4.

Zoccarato, F, L Cavallini, M Valente, and A Alexandre. 1999. "Modulation of Glutamate Exocytosis by Redox Changes of Superficial Thiol Groups in Rat Cerebrocortical Synaptosomes." *Neuroscience Letters* 274 (2): 107–10. http://www.ncbi.nlm.nih.gov/pubmed/10553949.

Zoccarato, F, M Valente, and A Alexandre. 1995. "Hydrogen Peroxide Induces a Long-Lasting Inhibition of the Ca(2+)-Dependent Glutamate Release in Cerebrocortical Synaptosomes without Interfering with Cytosolic Ca^{2+}." *Journal of Neurochemistry* 64 (6): 2552–58. http://www.ncbi.nlm.nih.gov/pubmed/7760035.

Chapter 2

Catalase in Littoral Macroalgae of the Barents Sea: Daily and Seasonal Changes in Activity, the Influence of Abiotic Factors

I. Ryzhik*, PhD and M. Makarov

Murmansk Marine Biological Institute of the Russian Academy of Sciences, Murmansk, Russia

Abstract

The enzyme catalase is an important component of the antioxidant system of algae. Its activity depends on their physiological state and on environmental factors. The catalase activity depends on the age of the thallus: in brown macroalgae *Fucus vesiculosus* the maximum is peculiar for the young part of the thallus (apical and middle), minimal observe in basal part. The presence of strong (2-4 times) fluctuations in catalase activity in *F. vesiculosus* during the day, which are synchronized with the tidal cycle, is shown. The main increases occur in the middle of low tide, as well as during periods when there is a change of habitat (transition from water to air and back). Especially significant changes are observed with a significant temperature/light gradient. In addition to daily changes in catalase activity, seasonal changes are peculiar also. During the year, the minimum catalase activity in *F. vesiculosus*, is observed in April, the maximum in January. During the summer and autumn, the activity of catalase practically does not change. In the red alga *Palmaria palmata*, the maximum catalase activity is observed in winter (December-February), the minimum in summer (July). Both types of algae have a similar tendency in changing catalase activity by seasons: two periods

* Corresponding Author's Email: alaria@yandex.ru.

In: Catalase and Its Applications
Editor: Kaley Rutherford
ISBN: 979-8-88697-421-8
© 2022 Nova Science Publishers, Inc.

can be distinguished, spring and autumn, associated with the adaptation of biochemical processes to changes in ambient temperature. In the spring, when the temperature and PAR increase, catalase inactivation occurs. In autumn, with a decrease in ambient temperature and PAR, catalase activity increases. During the period of decrease in catalase activity, its role can be performed by a complex of peroxidases. Catalase activity increases in algae (*F. vesiculosus*) in the presence of an petroleum products in the environment. Experimental studies have shown that catalase, along with other AOS enzymes, is activated in the first hours of the action of the toxicant (diesel fuel), and remains at a high level until the algae adapt. In general, the enzyme catalase is a sensitive component of the AOS of algae, which quickly reacts to changes in environmental factors and participates in increasing the adaptation of the organism.

Keywords: enzyme catalase, daily and seasonal changes, algae, *Fucus vesiculosus*, *Palmaria palmata*, Barents Sea

Introduction

The littoral (intertidal) zone of the seas is a harsh habitat for living organisms. The major factor in the littoral zone is the tidal cycle which determines the activity of other abiotic factors: light intensity, ambient temperature, salinity, availability of gases and biogens, velocity of interchange with the environment, etc. A very significant factor is also drainage at low tide, during which some species of algae can lose up to 50-70% of tissue moisture (Bisson, Kirst, 1995; Schagerl, Moostl, 2011, Ryzhik et al., 2021), reduce significantly the rate of photosynthetic and respiratory gas exchange (Quadir et al., 1979; Dring, Brown, 1982; Andreev et al., 2012; Lobban et al., 1985), etc. Many adaptations at different levels of organization, from morphological (changes in the shape and thallus structure and involucres) to physiological and biochemical (changes in growth rate, intensity of photosynthetic and respiratory processes, biochemical composition, etc.) occur in the plants for their existence of the intertidal zone (Gerard, 1982. Gao, Ji, Aruga, 1999; Flores-Molina et al., 2014).

In addition, the littoral can be considered as a zone between land and sea where a large amount of pollutants can accumulate, being brought both from land by winds and by waves from the sea to the coast. Therefore, the organisms living here can serve as promising bioindicator species that reveal the presence of various pollutants. Macroalgae are the most suitable flora for this purpose, as they are, as a rule, the most common coastal life forms.

An indicator of the presence of pollutants or significant changes in abiotic/biotic environmental factors may be a change in the physiological and biochemical parameters of the organism. This response may indicate the activation of its protective or adaptive mechanisms (Geret et al., 2003; Dring, 2006; Shakhmatova, Milchakova, 2014; Mallick 2004; Inupakutika et al., 2016).

One of the adaptive mechanisms is the activation of the antioxidant system which neutralizes reactive oxygen species (ROS) which are intensively formed in the cells of living organisms under stressful conditions. ROS are able to trigger a cascade of reactions leading to the destruction of cell membranes and, subsequently, to the death of the organism (Apel, Hirt, 2004 Karuppanapandian et al., 2011; Kreslavski et al., 2012; Waszczak et al., 2018 Kolupaev, 2007; Kolupaev, Karpets, 2010).

The components of the antioxidant system (AOS) are divided into two groups: antioxidant enzymes (catalase (CAT), superoxide dismutase, different types of peroxidases, etc.) and non-enzymatic antioxidants (for example, polyphenols, carotenoids, separate amino acids, etc.). The most studied enzymes are catalase, superoxide dismutase, glutathione peroxidase and ascorbate peroxidase. Also, the determination of hydrogen peroxide concentration in cells and lipid peroxidation (LPO) level are often used as a stress marker (Rogozhin et al., 2001; Rogozhin, 2004; Cantrell et al., 2003; Vega-Lopez et al., 2013; Pokora, Tukaj, 2010; Alscher et al., 2002, Migdal, Serres, 2011).

The catalase enzyme, in comparison with other components of the AOS of marine hydrobionts, is characterized by the widest range of response to changes in abiotic and biotic environmental factors (Shakhmatova, Milchakova, 2009, 2014; Kreslavski et al., 2012). Under normal conditions, the activity of this enzyme depends on the type of organism and its physiological features. Catalase, being an enzyme of the antioxidant system, neutralizes ROS, in particular, hydrogen peroxide. ROS are supposed to function as secondary messengers involved in inducing plant resistance to changes in abiotic factors (Davison, Pearson, 1996; Collén, Davison, 1999; Kreslavski et al., 2012).

The objective of this work was to analyze the catalase enzyme activity in different species of macroalgae in the Barents Sea, the annual and daily rhythms of the enzyme activity and the influence of various factors on the catalase activity.

Method

During this study, macroalgae sampled from their natural habitats (Zelenetskaya Bay, 69° 07.030′ N, 36° 04.228′ E and Kola Bay, Abram Cape 68°58.7034′ N. 33°1.1556′ E) were investigated. The experiments were conducted in the laboratory environment, in a thermal control room, at a temperature and light intensity similar to natural conditions. The work was carried out at the seasonal biological station of Murmansk Marine Biological Institute of the Russian Academy of Sciences (settlement of Dalniye Zelentsy) and in the main building of the institute (city of Murmansk) from 2012 to 2021.

Samples were taken at the same time (August) in order to compare the catalase activity in different systematic groups of macroalgae of the littoral zone.

For the study of the daily rhythms of catalase activity, the apical parts (1.5 cm) of even-aged thalli of *Fucus vesiculosus* Linnaeus, 1753 (5-7 dichotomous branches, length of 15-20 cm, weight of 20-30 g) growing in the littoral zone of the Zelenetskaya Bay were sampled. The material was collected every two hours during a 24-hour period in July of 2020.

To study the seasonal rhythms of catalase activity, brown algae *Fucus vesiculosus* and red *Palmaria palmata* were collected in the littoral zone near the settlement of Abram Cape of the Kola Bay. Collection was performed monthly at low tide in the first half of the day from 2015 to 2019. The enzyme activity in fucus algae, which have a complex thallus organization, was analyzed in its various parts (apex, middle part and stipe).

Fucus vesiculosus, *Palmaria palmata* and green alga *Ulvaria obscura* were used to determine the enzyme response to the presence of a toxicant (diesel fuel) in the habitat. Experimental studies were conducted in the laboratory environment. Algae were preliminarily acclimated to the laboratory environment for seven days and nights. Both the influence of the water-accommodated fraction of hydrocarbons (WAF) and the film of oil products were measured.

To determine the effect of WAF, diesel fuel thoroughly mixed with sea water at a concentration of 0.8 μg/l was added to vessels with acclimated algae. Algae in other vessels kept in pure sea water were considered to be the control group. The catalase activity in algae both in the experiment and in the control was determined after 14 days. The apical part of the thallus of *Fucus vesiculosus* up to 0.5 cm long was used for the analysis. For *Palmaria palmata* and *Ulvaria obscura*, the entire thallus was used.

In order to determine the effect of a film of oil products on the surface of previously acclimated thalli of *Fucus vesiculosus*, an oil film was precipitated. To do this, crude oil was added to sea water at a rate of 2 ml/l of water, and the resulting mixture was poured through a Buchner funnel into which algae were placed. As a result, a film was formed on the thallus surface. Algae treated the same except with clean sea water served as a control. Then the algae were placed in a cuvette in the air for 5 hours. This corresponded to the presence of algae in the natural environment on the littoral at low tide. CAT activity was measured in 30 minutes and then every hour.

To analyze the daily and annual rhythms of CAT activity in algae sampled from their natural habitats, algal thalli were either fixed with liquid nitrogen or treated within 2 hours after sampling, depending on the distance of the sampling site from the place of their laboratory analysis. When sampling, the air and water temperatures were measured with a TL-4 mercury thermometer (TU 25-2021.003-88, Russia, accuracy 0.01), the PAR intensity was measured with a LI-185A photometer (LI-COR, Lambda Inst., Nebraska, USA) and the stage of a tidal cycle (TC) and the water level was measured with a depth indicator.

Catalase Activity

Catalase activity was determined according to the modified spectrophotometric method based upon the ability of hydrogen peroxide to form a stable colored complex with molybdenum salts (Korolyuk et al., 1988). Alga sample (150-200 mg wet weight) was ground on ice in a mortar with 2 ml of K/Na-phosphate buffer (pH 7.8) added. The homogenate was centrifuged at 4°C at 8000 g for 5 minutes. Two ml of 0.03% hydrogen peroxide solution was added to 0.1 ml of supernatant. One-tenth ml of distilled water was added to the blank sample instead of supernatant. The reaction was carried out at a temperature of 18°C and was stopped in 10 minutes by adding 1 ml of 4% ammonium molybdate. Colour intensity was measured with the JENWAY 6305 UV/VIS Spectrophotometer (Jenway, Essex, UK) at a wavelength of 410 nm. CAT activity was defined as the difference in the final colour intensity of the sample compared to the blank. The enzyme activity was calculated on the basis of the dry weight. The measurements were taken three times and the values averaged.

Data Analysis

The obtained data were processed and analyzed using the statistical software package of Microsoft Excel 2010.

Result and Discussion

CAT activity in macroalgae of various systematic groups was determined at the first stage of the study during summer. The maximum enzyme activity was observed in the red alga *Palmaria palmata* and the minimum in the brown alga *Ascophyllum nodosum* out of the seven studied species. CAT activity in other algal species was similar (Figure 1).

Differences in the enzyme activity in different algal species can be principally conditioned by their physiological features, structure and stage in the life cycle.

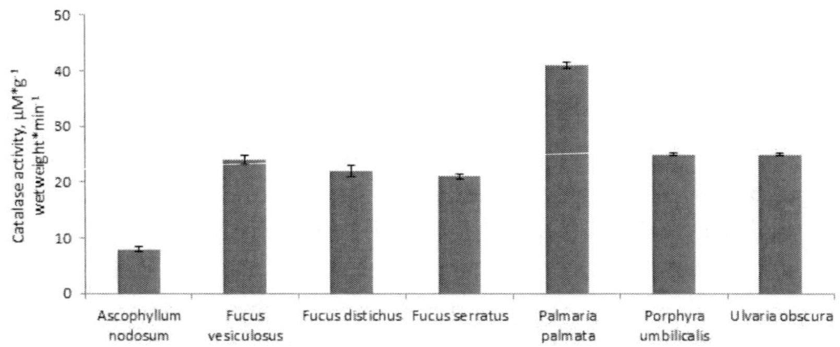

Figure 1. CAT activity in different macroalgal species in the summer period (August of 2015, Zelenetskaya Bay in the Barents Sea). The data is presented as an arithmetic mean and standard deviation.

The thallus of Fucus algae has a complex structure. There is a cuticular layer on its surface which prevents the evaporation of moisture from the surface and thus reduces the loss of water by algal tissues at low tide when the algae are in the air. Drying and the associated change in abiotic environmental factors (temperature, light intensity, salinity, etc.) is a stress for littoral algae which leads to the ROS accumulation and the activation of antioxidant system. In addition to catalase, there are other enzymes in fucus algae which perform an antioxidant function. These are peroxidase, ascorbate peroxidase, superoxide dismutase, etc., which form part of the enzyme component of AOS. In addition to enzymes, fucoids contain additional compounds, such as carotenoids, polyphenols, free amino acids, which are also capable of performing an antioxidant function. Perhaps this abundance of antioxidants is why the CAT activity in fucus algae is not so high. The enzyme activity in the red alga *Palmaria palmata* is much higher. It is possible that the complex of other antioxidant enzymes and substances is absent or insufficiently active in this species, and catalase plays the role of the main antioxidant.

CAT Activity in Different Parts of *Fucus vesiculosus* Thallus

Fucus algae are perennial and have a dichotomously branched complex thallus. Fucus thalli can be conditionally divided into a stipe (lower part with an adhesive organ, i.e., a sole), a middle part and an apical part. The stipe is the oldest part of the thallus, the apical part is the youngest and actively growing. Analysis of CAT activity was different in different parts of the thalli. During the winter, the greatest activity was observed in the young apical parts whereas the least was observed in the stipe. In springtime, when an active growth of algae and intensification of metabolism are observed, CAT activity in the apical and middle parts is equalized (Figure 2).

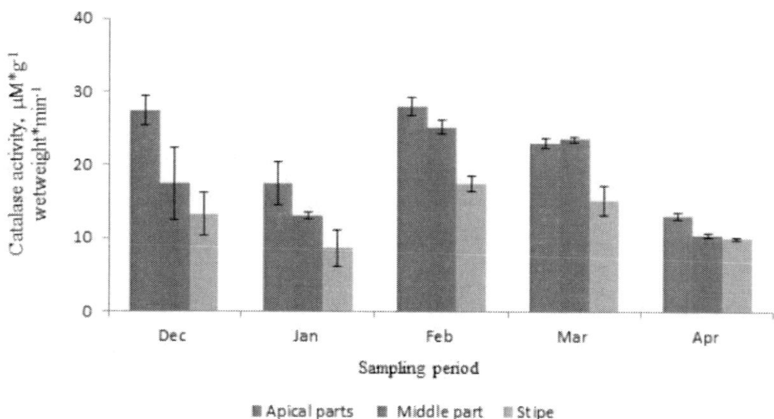

Figure 2. CAT activity in different parts of *Fucus vesiculosus* thallus, winter and spring period. The data is presented as an arithmetic mean and standard deviation.

Most likely, the differences are due to the functional and physiological features of the anatomical parts under consideration: cell division and growth occur in the apical parts, and metabolic activity is maximum there (Moss, 1967). Photosynthetic rate and respiration intensity are the highest in the middle parts of the thalli. A large number of metabolites are synthesized there, which are then transported to the rest parts of the thallus (Schmitz, Lobban, 1976; Diouris, Floc'h, 1984; Kamnev, 1989). Accordingly, the work of the antioxidant system in these parts is the most intensive. The stipe possesses the least physiological activity and is mainly responsible for a mechanical function and for accumulation of reserve nutrients (Lüning et al., 1973; Schmitz, Srivastava, 1979; Diouris, 1989; Wheeler, North, 1981; Birkemeyer et al., 2019).

An increase in metabolic activity in a certain zone will cause ROS accumulation and, as a result, an increase in the functioning of antioxidant system. Metabolic activity depends on the season and the "life strategy" of the organism in a particular period of the year. In winter, receptacles are formed and develop in the apical parts of fucus algae, which are the reproductive organs of this algal species. And during this very period, the CAT activity in the apex is maximum. In February, with an increase in the length of daylight hours and light intensity, active growth processes begin to occur. During this period, the photosynthetic apparatus of the middle part of the thallus and the division and growth of cells in the apical part are activated. These are accompanied by an increase in the metabolic activity of these parts and ROS accumulation in their cells. In this regard, CAT activity increases in the apex and in the middle part in February-March (Figure 2).

Circadian Rhythms of Catalase Activity

Algae, being living organisms, are characterized by the presence of circadian rhythms of various physiological and biochemical parameters, with a periods from several hours up to a year. These cycles regulate cell division and growth, photosynthetic rate, organelle migration in the cell, gene expression, etc. (Makarov et al., 1995; 1999; Goulard et al., 2004; Titlyanov et al., 2006; Gupta, Kushwaha, 2017; Stepanyan, 2019). Cycles are related to periodic changes in abiotic environmental factors (PAR, temperature, nutrient concentration, etc.). Tidal cycles are also an important factor for littoral algae which determine the periodic exposure of algae to the air (Thomas, Turpin, Harrison, 1987; Bischof et al., 2006; Karsten, 2012; Mullineaux et al., 2018, Diehla et al., 2019).

The daily changes in various physiological processes are based on biorhythms which are synchronized with the cycles in nature, namely, high tides and low tides and alternation of day and night. They can be attributed to adaptive (ecological), genetically fixed ones. Biological rhythms in photosynthetic organisms are complex and most likely regulated by several mechanisms (Hastings et al., 2001. Falcão et al., 2010).

Circadian Rhythms

In the summer period (July 2019), strong (2-4 times) fluctuations in catalase activity in the brown alga *Fucus vesiculosus* were shown during the day with a period of about 4 hours.

The main increases in the enzyme activity occur in the middle of low tides, as well as during the periods when a change in habitat takes place (transition from water to air and vice versa). Particularly significant changes occur in response to a significant temperature and light gradient, for example, on a hot and windy day. Under such conditions, the temperature difference between sea water and air can reach 20°C, and when coastal waters are stirred up, the intensity of photosynthetically active radiation penetrating into the water decreases sharply. During the tidal cycle, this is repeated in about 6 hours.

Changes in CAT activity indicate that ROS concentration in algae cells during the day and night is not stable. This is indicative of a change in the activity and orientation of physiological processes conditioned, among other things, by periodic changes in environmental factors. Daily fluctuations of physiological parameters have been observed for many plants. An increase in ROS concentration and in LPO level is observed in *Ulva lactuca* at low tide and stabilization at high tide (Gupta, Kushwaha, 2017). Daily dynamics of another antioxidant enzyme activity, superoxide dismutase, was revealed in microalgal cell cultures, i.e., an increase in activity during the day and a decrease at night. At the same time, the circadian rhythm of enzyme activity was maintained even under continuous illumination (Barros, 2005; Rossa et al., 2002). The daily dynamics of accumulation of ROSs, which perform a signaling function and participate in the regulation of plant metabolism, is also revealed in higher plants (Li et al., 2020). An increase in the amount of ROSs and LPO products can activate the transition to another metabolic pathway and lead to a change in dominant processes, i.e., nitrogen and carbohydrate metabolism and carboxylation type in photosynthetic processes (Ryzhik et al., 2021).

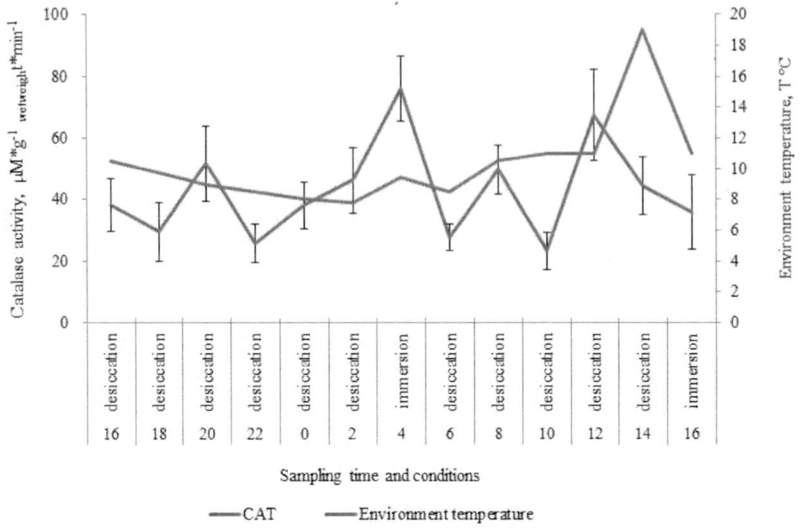

Figure 3. Circadian rhythms of catalase activity in *Fucus vesiculosus* (July 2019). The data is presented as an arithmetic mean and standard deviation.

Annual Rhythms

During the year, CAT activity was determined in two macroalgal species belonging to different systematic groups, having different structures, but inhabiting the study area under approximately the same conditions. These are brown algae *Fucus vesiculosus* and red *Palmaria palmata.* In order to minimize the effect of diurnal rhythms, samples were collected in the first phase of low tide, at about 10–12 AM, once a month.

The results of the studies showed that during the year, the studied algal species have several periods of significant changes in CAT activity, timed to the seasons and the species' "life strategy" (Figure 4).

The seasonal rhythms of CAT activity with a period of 3–4 months can be observed most clearly in *F. vesiculosus*. Thalli of this species develop receptacles (reproductive organs) in December. The active formation of reproductive products during this period increases cellular metabolic activity, resulting in ROS accumulation and intensified functioning of the antioxidant system. The next peak of CAT activity occurred during February, the month when daylight hours and PAR intensity increase. At this time, algae begin to grow, i.e., cell division and growth occur, which is accompanied by an increase in metabolic activity. The summer peak occurred during June, when the polar day begins, the temperature rises, maximum insolation is observed, and algal growth rate is maximum (Makarov, 1999). All this leads to ROS accumulation in cells. During this period, the intensity of metabolic processes increases sharply (Ryzhik, 2016), while CAT activity increases 2.2 times. The autumn peak is associated with the active synthesis of reserve nutrients and their transport through the thallus, which is accompanied by an increase in energy metabolism and, accordingly, by ROS accumulation again.

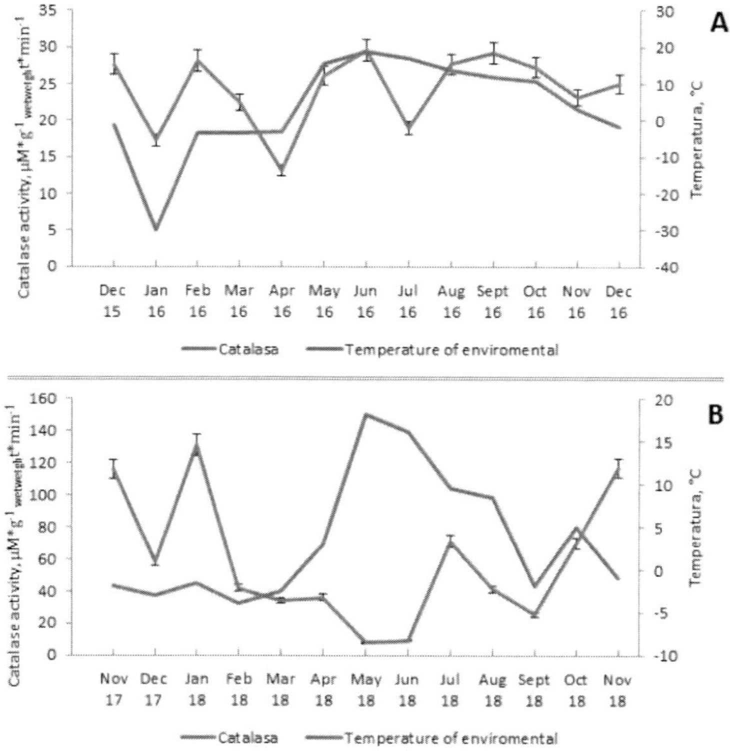

Figure 4. Catalase activity in *Fucus vesiculosus* (A) and *Palmaria palmata* (B) during the year. The data is presented as an arithmetic mean and standard deviation.

The differences between the maximum and minimum CAT activity were 2-3 fold in *F. vesiculosus* and 3-4 fold in *P. palmata*. Reproduction in this species occurs in January and is accompanied by an increase in CAT activity. There is also a June peak related to temperature and light conditions. However, the active synthesis of reserve nutrients and the associated increase in CAT activity do not occur during September, as in *F. vesiculosus*, but rather in November.

In the spring period, with an increase in light intensity and duration, a decrease in CAT activity was observed in both species. It is possible that this is related to the rearrangement of the photosynthetic apparatus and to an increase in the relative content of carotenoids which are also able to neutralize free radicals (Makarov, 2012). During this period, polyphenolic compounds, the maximum accumulation of which is observed in March, also play a significant role in protecting macroalgal cells against ROS (Ryzhik, Fisak, 2018). In general, the relationship between the antioxidant enzyme activity level and resistance to photooxidative stress was noted in both brown and red macroalgae of the North Atlantic (Collen, Davidson, 1999).

Little variation in CAT activity in *F. vesiculosus* occurred from August to October. This is indicative of the intensive work of AOS components, which is also confirmed by the high cellular metabolic activity, especially in July and August (Ryzhik, 2016). These findings are consistent with other observations of the high resistance of *F. vesiculosus* to high PAR intensity and UV radiation compared to other macroalgal species in the Barents Sea (Makarov 1999).

In general, an increase in CAT activity during the preparation of macroalgae for winter, compared to the summer period, was described heretofore (Willekens et al., 1995; Shakhmatova, Milchakova, 2009). These data are also consistent with the studies performed by Belotsitsenko (2015) which, with a decrease in temperature, showed the carotenoids accumulation and an increase in the activity of antioxidant enzymes in dominant macroalgae in the Sea of Japan.

Changes in Catalase Activity at Low Tide upon Algae Drying

As already mentioned for circadian rhythms, CAT activity is to a large extent determined by environmental factors. Drying is one of the main factors for macroalgae growing in the littoral area. We found that CAT activity decreased in *F. vesiculosus* during the first hours of low tide, then it sharply increased (almost 2 times). By the end of low tide, activity decreases again (1.5 times) (Figure 5). This study was conducted during four periods of drying and the nature of change was similar in each period. A certain time shift of the peaks in either direction depending on weather conditions (temperature, light intensity, wind speed, humidity) were noted, i.e., on hot days, periods of decrease and increase in enzyme activity occurred sooner, which probably shows the dependence of CAT activity on the rate of moisture loss by the algal thalli during the drying period.

Changes in AOS enzyme activity were also noted for other algal species. For example, a significant increase in ROS level was observed upon drying in seaweed *Ulva lactuca*, a member of the genera Pyropia and Porphyra. In this case, LPO products accumulate, which, in turn, triggers the synthesis or activation of AOS enzymes responsible for their inactivation (Gupta et al., 2017). Fluctuations in superoxide dismutase activity are also conditioned by the peculiarities of light regime, such as light intensity, qualitative composition of light, etc. (Wiencke, Gómez, Dunton, 2009; Barros et al., 2005). Catalase is characterized by more complex circadian rhythms, since it performs not only the peroxidase function, but also participates in other processes (Willekens et al., 1995 Kolupaev, Karpets, 2010).

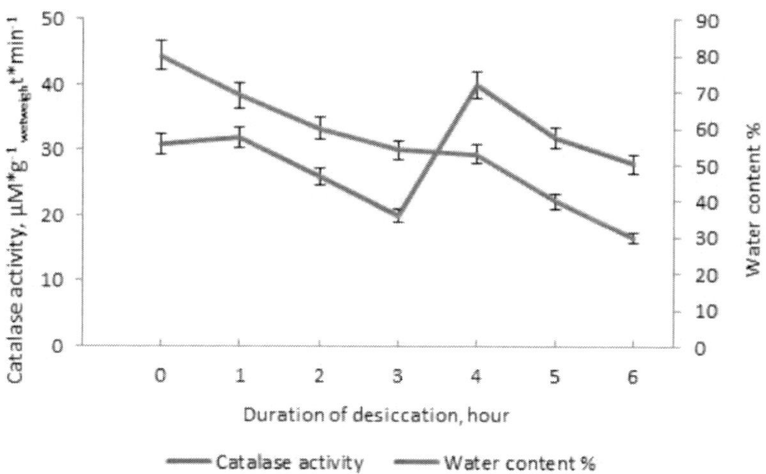

Figure 5. Catalase enzyme activity in the cells of *F. vesiculosus* at low tide. The data is presented as an arithmetic mean and standard deviation.

Changes in Catalase Activity When the Marine Environment Is Polluted with Oil Products

When the marine environment is polluted with oil products as a consequence of accidental spills, the most vulnerable are the inhabitants of the littoral area, These species are affected not only by water-soluble or emulsified fractions of hydrocarbons, but also by the film of oil products deposited on them at low tide. They are further subjected to significant fluctuations in the environmental factors during tidal cycles (Quadir et al., 1979; Dring, Brown, 1982; Bisson, Kirst, 1995; Schagerl, Moostl, 2011; Andreyev et al., 2012). As a rule, algae are more resistant to the impact of oil products, especially those with a well-developed cuticular layer on the thallus surface. This cuticular layer slows the penetration of oil products into the cellular underlayers and, thus, protects the plant against the harmful effect of the toxicant (Filion-Myklebust, Norton, 1981). Also, algae in which regeneration processes occur more intensively and/or defense mechanisms are activated will be more resistant, for example, inactivation of toxicants by reactive oxygen species, etc.

Impact of an Oil Product Film on CAT Activity in *Fucus vesiculosus*

During the experiment, oil film deposition on the surface of *F. vesiculosus* at low tide was simulated. After a film had deposited on them, algae were left in the air for 5 hours. Algae without oil treatment and kept under the same conditions served as the control. The results showed that CAT activity in the control and experimental plants did not differ during the first two hours following exposure to a film of oil products. During the third hour of the experiment, CAT activity in the experimental plants became higher than that of the control plants, but then it decreased to the control values. In case of algae transfer to the aquatic environment (tide simulation), CAT activity in both groups increased (Figure 6).

The negative effect of the film of oil products is most likely related to the direct impact of the toxicant on the protein-lipid complex of membranes. Additionally, the film affects the gas exchange between algae and the environment, inhibits the evaporation of moisture from the thallus surface which causes an increase in its temperature and changes the intensity and spectral composition of the light that reaches the photosynthetic apparatus of the thallus cells. When exposed to an oil film, the response of the algal antioxidant system appears within 3 hours. The two-hour duration of the period of nonresponse to the oil film on the algal surface is indicative of the barrier function of the cuticular layer which prevents the toxicant from penetrating into the physiologically active layers of the thallus cells. It is possible that the cumulative effect of the film impact manifests itself in 3 hours (an increase in the thallus temperature, a decrease in the possibility of gas exchange, the toxic effect of petroleum products, etc.), which leads to a stress response of the organism and an increase in ROS level in the algal cells.

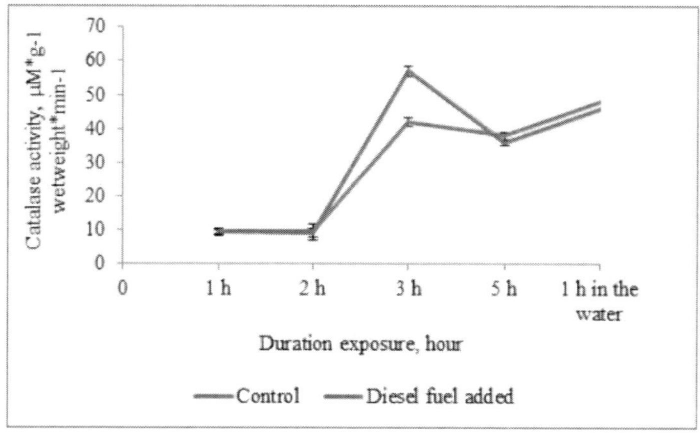

Figure 6. Catalase activity in *Fucus vesiculosus* following an oil film treatment.

Impact of Emulsified Petroleum Hydrocarbons on CAT Activity in *Fucus vesiculosus*

The results of the experiment on the impact of diesel fuel-water emulsion showed that CAT activity changes only during the first 24 hours, i.e., a 1.5-fold decrease in the enzyme activity was observed compared to the control. Between 24 and 48 hours, CAT activity increased and for 6 days it remained insignificantly higher or similar to the control (Figure 7).

It is unclear what the reason for the decrease in CAT activity in fucoids was during the first day. It is possible that after the activation of the enzyme component of the antioxidant system during the first hours of exposure, other components of the antioxidant system were subsequently involved. They were carotenoids, free amino acids, etc. which required a certain period of time for their synthesis. And the need for "fast" enzymatic responses to ROS accumulation decreased with the beginning of their functioning.

A number of conducted experiments have shown a similar response of catalase in *Fucus vesiculosus*, *Acrosyphonia arcta* to emulsified types of petroleum hydrocarbons. At the same time, a decrease in CAT activity in all species during the first day and its further stabilization were observed regardless of the concentration of introduced oil products (the studies were carried out with diesel fuel concentrations 0.86 - 129 mg/l) (Ryzhik et al., 2022).

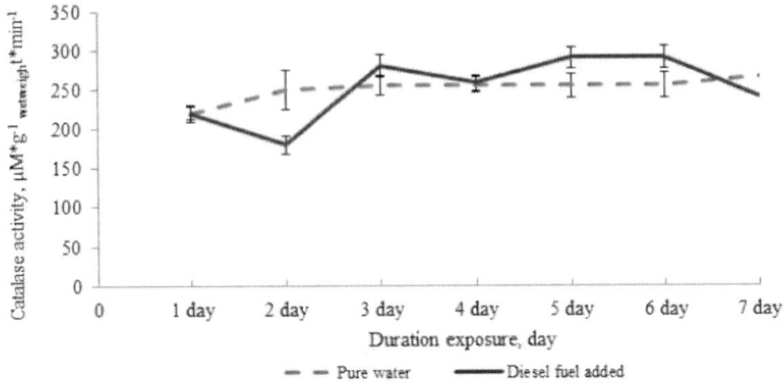

Figure 7. Catalase activity in the cells of *F. vesiculosus* during the experiment. The data is presented as an arithmetic mean and standard deviation.

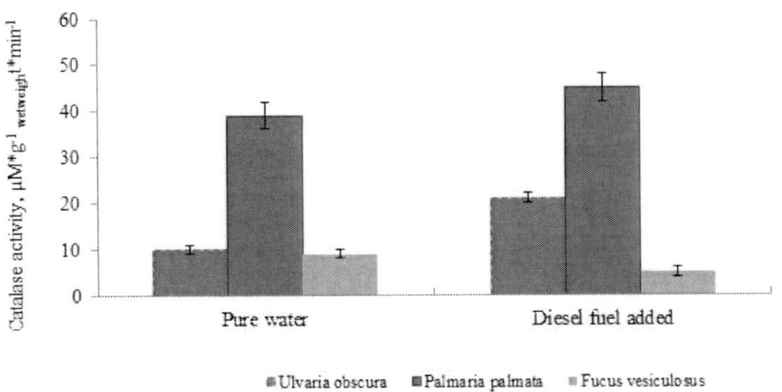

Figure 8. Changes in catalase activity in algae under the impact of diesel fuel- water emulsion at a concentration of 0.8 mg/l. The data is presented as an arithmetic mean and standard deviation

It is of interest to note that long-term (7 days) experiments on the effect of emulsified petroleum hydrocarbons on CAT activity in various algal species showed that brown algae *F. vesiculosus* and red *Palmaria palmata* have the same response. In contrast, green U. obscura reacted differently as at the end of the experiment, CAT activity was 2 times higher compared to the control (Figure 8). Probably, this is primarily related to the algae structure. As distinct from *F. vesiculosus* and *P. palmata*, the thallus of U. obscura consists of one row of cells without a pronounced cuticular layer, and therefore the penetration of toxicants into the cells occurs more readily. In this case, it is possible that the non-enzymatic component of the antioxidant system cannot manage it and the enzymatic component functions for a long time.

Conclusion

ROS formation in photosynthetic organisms is an obligatory process which accompanies various metabolic processes, namely, photosynthesis, respiration, nitrogen fixation, etc. (Vardi et al. 1999). Therefore, the synthesis of enzymatic and non-enzymatic compounds forming the antioxidant system occurs constantly. The intensity of the synthesis of the antioxidant system enzymes is a species-specific feature. In different seasons of the year, synthesis intensification and enzymatic activation are associated with the species' "life strategy," i.e., the processes accompanied by an increase in the cellular metabolic activity. These are the growth of thalli in the spring and summer period, the synthesis, transport and accumulation of nutrients in autumn, the formation of reproductive organs and products in the autumn and winter period, etc. (Ryzhik, 2016). Except for the internal characteristics of the organism, the enzymatic activity is influenced by external abiotic and biotic factors. Their rapid change causes stress in the organism which is accompanied by an increase in the concentration of reactive oxygen species in the cells. This, in turn, causes the activation of enzymatic and non-enzymatic components of the antioxidant system. High concentrations of antioxidants will be crucial for the regulation of photo-oxidative stress caused by a decrease in energy consumption due to the action of xenobiotics (Okamoto et al. 2001).

Our studies of green algae species have shown that the increased CAT activity persists for a long time when they are in an environment polluted with oil products. In brown and red algae, the activities of the AOS enzymatic component in them are quickly substituted for non-enzymatic components. Such differences in the response of enzymes to the effect of a xenobiotic can be attributed to a number of reasons, namely, toxicant nature, species of a plant, stage of its life cycle, etc. (Torres et al., 2008; Pilatti et al., 2006, 2017; El-Shoubaky, Mohammad, 2016; Bokn et al., 1993). Thus, early stages of development (germlings, zygotes, gametes) are more sensitive than full-grown plants, since the latter are able to adapt to such an impact. Full-grown plants also adapt by means of AOS activation, with the help of the synthesis of heat shock proteins, etc. Studies of oil spills at sea during which dispersing agents were used for clean up have shown that, thanks to the synthesis of heat shock proteins in algae, their resistance to crude oil increases (Wolfe et al. 1998). Likewise, some concentrations of petroleum products can have a stimulating effect on the organism (Lewis, Pryor, 2013).

The degree of toxicity for the algal organism can also be influenced by its structural features. Fucus algae are characterized by a complex tissue structure with cell differentiation as a dense cuticular layer is formed on the thallus surface (Kamnev, 1989). Various bacterial species develop on it, including those capable of oxidizing petroleum products (Pugovkin et al. 2016). Presumably, the penetration of oil products to the underlayers slows down precisely due to the presence of this layer and to the activity of bacteria which neutralize adsorbed oil products (Grande, Reis, Jacobucci, 2012). U. obscura is characterized by a single-layer plate which does not have a protective layer. As such, when an oil product appears on the thallus surface, it quickly penetrates into the cells. Oil penetration rate into *P. palmata* is intermediate between fucus algae and *U. obscura*, i.e., it has several cell layers and there is a thin cuticular layer on the surface which slows down the penetration of the toxicant.

Comparison of CAT activity in *F. vesiculosus* thalli growing in the areas with different oil pollution levels (with concentrations from 0.023 to 0.057 mg/dm^3) showed no differences in enzyme activity. The same nonresponse to pollution was noted in the allied species *Cystoseira crinita* growing in the Black Sea (Shakhmatova, Milchakova, 2009). It is possible that, in case of chronic oil pollution, another system responsible for ROS neutralization is activated in fucus algae. This may be ascorbate peroxidase which reduces peroxides to H_2O or the corresponding alcohols using ascorbate as an electron donor (Torres et al., 2008). Also, other compounds such as carotenoids, polyphenols, amino acids, etc. can act as long-acting antioxidants.

Thus, the level of catalase activity in macroalgal cells is not a stable indicator, but has pronounced daily and annual rhythms associated with the tidal cycle, seasonal activity of algae and environmental conditions.

Disclaimer

None.

References

Alscher, Ruth G., Neval Erturk and Lenwood S. Heath. 2002. "Role of superoxide dismutases (SODs) in controlling oxidative stress in plants." *Journal of experimental botany* 53 372: 1331-41.

Andreev, V. P., Maslov, Yu. I., and Sorokoletova, E. F. 2012. "Functional properties of photosynthetic apparatus in three fucus species inhabiting the White Sea: effect of dehydration." *Russ. J. Plant Physiol* 59 2: 217–223.

Apel K., Hirt H. 2004. "Reactive oxygen species: Metabolism, Oxidative Stress, and Signal Transduction." *Annual Review Plant Biology* 55: 373–99 doi: 10.1146/annurev.arplant.55.031903.141701.

Barros M. P, Ernani P., Sigaud-Kutner T. C. S, Cardozo K. H. M, Dr Colepicolo P. 2005. "Rhythmicity and oxidative/nitrosative stress in algae." *Biological Rhythm Research* 36 1-2: 67-82, doi: 10.1080/09291010400028666.

Belotsitsenko, E. S. 2015. "*Resistance of marine macroalgae to photo-oxidative stress under conditions of temperature fluctuation*" PhD diss, –Vladivostok: FEB RAS, IBM. [In Russia].

Birkemeyer C., Osmolovskaya N., Kuchaeva L., Tarakhovskaya E. 2019."Distribution of natural ingredients suggests a complex network of metabolic transport between source and sink tissues in the brown alga *Fucus vesiculosus*." *Planta* 249 2: 377-391. doi: 10.1007/s00425-018-3009-4.

Bischof K., Gómez I., Molis M., Hanelt D., Karsten U., Lüder U., Roleda M. Y., Zacher K., Wiencke C. 2006. "Ultraviolet radiation shapes seaweed communities." *Rev. Environ. Sci. Bio. Technology* 5: 141–166. https://doi.org/10.1007/s11157-006-0002-3.

Bisson M. A. and Kirst G. O. 2005. "Osmotic acclimation and turgor pressure regulation in algae." *Naturwissenschaften* 82: 461-471.

Bokn T. L., Moy R. E., Murray S. N. 1993. "Long-term Effects of the Water-accomodated Fraction (WAF) of Diesel Oil on Rocky Shore Populations" *Botanica Marina* 36: 313-319.

Cantrell A., McGarvey D. J., Truscott T. G., Rancan F., BoËhm F. 2003. "Singlet oxygen quenching by dietary carotenoids in a model membrane environment." *Archives of Biochemistry and Biophysics* 412: 47–54.

Collén J. and Davison I. R. 1999."Stress tolerance and reactive oxygen metabolism in the intertidal red seaweeds *Mastocarpus stellatus* and *Chondrus crispus*" *Plant, Cell Environ* 22: 1143–1151.

Davison I. R. and Pearson G. A. 1996. "Stress tolerance in intertidal seaweeds" *J. J. Phycol.* 32: 197-211.

Diehl N., Michalik D., Zuccarello G. C., Karsten U. 2019 "Stress metabolite pattern in the eulittoral red alga *Pyropia plicata* (Bangiales) in New Zealand – mycosporine-like amino acids and heterosides." *Journal of Experimental Marine Biology and Ecology* 510: 23–30. https://doi.org/10.1016/j.jembe.2018.10.002.

Diouris M. 1989. "Long-distance transport of 14C-labelled assimilates in the Fucales: nature of translocated substances in *Fucus serratus*." *Phycologia*. 28(4): 504-511.

Diouris M., Floc'h J. Y. 1984. "Long-distance transport of C-labelled assimilates in the Fucales: directionality, pathway and velocity." *Marine Biology*. 78: 199-204.

Dring M. J., Brown F. A. 1982. "Photosynthesis of intertidal brown algae during and after periods of emersion: a renewed search for physiological causes of zonation" *Mar Ecol. Prog. Ser.* 8: 301-308.

Dring M. J. 2006. “Stress resistance and disease resistance in seaweeds: the role of reactive oxygen metabolism.” *Advances in botanical research*. 43: 175–207.

El-Shoubaky G. A., Mohammad S. H. 2016. “Bioaccumulation of gasoline in brackish green algae and popular clams.” *Egyptian Journal of Aquatic Research* 42: 91–98.

Falcão V. R., Oliveira C. M, Colepicolo P. 2010. “Molecular characterization of nitrate reductase gene and its expression in the marine red alga *Gracilaria tenuistipitata* (Rhodophyta).” *J Appl Phycol* 22:613–622. doi 10.1007/s10811-010-9501-2.

Filion-Myklebust, C. and T. A. Norton. 1981. “Epidermis shedding in the brown seaweed *Ascophyllum nodosum* (L.) Le Jol. and its ecological significance.” *Mar. Biol. Lett.* 2: 45–51.

Flores-Molina M. R., Thomas D., Lovazzano C., Núñez A., Zapata J., Kumar M., & *Contreras-Porcia*, L. 2014. “Desiccation stress in intertidal seaweeds: Effects on morphology, antioxidant responses and photosynthetic performance.” *Aquatic Botany*. 113, 90-99.

Gao K., Ji Y., Aruga Y. 1999. “Relationship of CO_2 concentrations to photosynthesis of intertidal macroalgae during emersion.” *Hydrobiologia*. 398:355–359. https://doi.org/10.1023/A:1017072303189.

Gerard V. A. 1982. “Growth and utilization of internal nitrogen reserves by the giant kelp *Macrocystis pyrifera* in a low-nitrogen environment.” *Marine Biology*. 66: 27–35. doi: 10.1007/BF00397251.

Geret F., Serafim A. & Bebianno, M. J. 2003. “Antioxidant Enzyme Activities, Metallothioneins and Lipid Peroxidation as Biomarkers in *Ruditapes decussatus*?” *Ecotoxicology* 12:417–426. https://doi.org/10.1023/A:1026108306755.

Goulard F., Lüning K., Jacobsen S. (2004). Circadian rhythm of photosynthesis and concurrent oscillations of transcript abundance of photosynthetic genes in the marine red alga *Grateloupia turuturu*. *European Journal of Phycology*. 39:4. 431-437. doi: 10.1080/09670260400009908.

Grande H., Reis M., Jacobucci G. 2012. “Small-scale experimental contamination with diesel oil does not affect the recolonization of Sargassum (Fucales) fronds by vagile macrofauna.” *Zoologia*. 29 2:135-143.

Gupta V. and Kushwaha H. R. 2017. “Metabolic regulatory oscillations in intertidal green seaweed *Ulva lactuca* against tidal cycles.” *Scientific reports*. 7:1-9. (16430). doi.org/10.1038/s41598-017-15994-2.

Hastings J. W. 2001. “Cellular and Molecular Mechanisms of Circadian Regulation in the Unicellular Dinoflagellate *Gonyaulax polyedra*. In: Takahashi J. S., Turek F. W., Moore R. Y. (eds) *Circadian Clocks. Handbook of Behavioral Neurobiology*. Vol. 12. Springer, Boston, MA. P. 321-334. https://doi.org/10.1007/978-1-4615-1201-1_12.

Inupakutika, Madhuri A., Soham Sengupta, Amith R. Devireddy, Rajeev K. Azad and Ron Mittler. 2016. “The evolution of reactive oxygen species metabolism.” *Journal of experimental botany* 67 21: 5933-5943.

Kamnev A. N. 1989. “*Structure and functions of brown algae.*” Moscow: MSU Publishing House, 200 p. ISBN 5-211-01449-9 [In Russia].

Karsten U. 2012 “Seaweed Acclimation to Salinity and Desiccation Stress.” Seaweed Biology. *Ecological Studies* (Analysis and Synthesis). 219.. https://doi.org/10.1007/978-3-642-28451-9_5.

Karuppanapandian, T., Moon, J. C., Kim, C., Manoharan, K., & Kim, W. 2011. "Reactive Oxygen Species in Plants: Their Generation, Signal Transduction, and Scavenging Mechanisms." *Australian Journal of Crop Science*. 5(6): 709–725. https://search.informit.org/doi/10.3316/informit.282079847301776.

Kolupaev Yu. E. 2007. "Active forms of oxygen in plants under the action of stressors: education and possible functions." *Bulletin of the Kharkiv National Agrarian University series biology*. 3 12: 6-26 [In Russia].

Kolupaev Yu. E., Karpets Yu. V. 2010. "*Formation of adaptive reactions of plants to the action of abiotic stressors.*" K: Osnova. 352 p. [In Russia].

Kreslavski V. D., Los D. A., Allakhverdiev S. I., & Kuznetsoy, VI. V.et al. 2012. "Signaling role of reactive oxygen species in plants under stress." *Russ J Plant Physiol* 59:141-154. https://doi.org/10.1134/S1021443712020057.

Lewis, M., Pryor, R. 2013. "Toxicities of oils, dispersants and dispersed oils to algae and aquatic plants: review and database value to resource sustainability." *Environmental Pollution*. 180: 345-367.

Li D., Wei J., Peng Z., Ma Wenna, Qian Yang, Zhongbang Song, Wei Sun, Wei Yang, Li Yuan, Xiaodong Xu, Wei Chang, Zed Rengel, Jianbo Shen, Russel J. Reiter, Xiuming Cui, Dashi Yu, Chen Q. 2020. "Daily rhythms of phytomelatonin signaling modulate diurnal stomatal closure via regulating reactive oxygen species dynamics in Arabidopsis." *J Pineal Res*.;68:e12640. https://doi.org/10.1111/jpi.12640.

Lobban C. S., Harrison P. J., Duncan M. J. 1985. "*The physiological ecology of seaweeds.*" New York, Cambridge University Press: 242 p.

Lüning K., Schmitz K., Willenbrink J. 1973. "CO_2 fixation and translocation in benthic marine algae. III. Rates and ecological significance of translocation in *Laminaria hyperborea* and *L. saccharina*." *Marine Biology*. 23(4): 275-281.

Makarov M. V. 2012. "Adaptation of the light-harvesting complex of the Barents sea brown seaweed *Fucus vesiculosus* L. to light conditions" *Doklady Biological Sciences*. 442(1): 58-61.

Makarov M. V. 1999. "Influence of ultraviolet radiation on the growth of the dominant macroalgae of the Barents Sea. In: *Chemosphere: Global Change Science*. "Climate Change Effect on Northern Terrestrial and Freshwater Ecosystems. 1 4: 461-467. https://doi.org/10.1016/S1465-9972(99)00034-3.

Makarov V. N., Makarov M. V., Schoschina E. V. 1999. "Seasonal dynamics of growth in the Barents Sea seaweeds: endogenous and exogenous regulation." *Botanica Marina*. 42 1:43-49.

Makarov V. N., Schoschina E. V., Lüning K. 1995. "Diurnal and circadian periodicity of mitosis and growth in marine macroalgae. I. Juvenile sporophytes of Laminariales (Phaeophyta)." *European Journal of Phycology*. 30 4: 261-266. doi: 10.1080/09670269500651031.

Mallick N. 2004. "Copper-induced oxidative stress in the chlorophycean microalga *Chlorella vulgaris*: response of the antioxidant system." *Journal of Plant Physiology*. 61 5: 591-597, https://doi.org/10.1078/0176-1617-01230.

Migdal C. and Serres M. 2011. "Reactive oxygen species and oxidative stress." *Medecine Sciences*: M/S. 27 4:405-412. DOI: 10.1051/medsci/2011274017. PMID: 21524406.

Moss B. 1967. "The apical meristem of Fucus." *New Phytologist*. 66 1:67-74. https://doi.org/10/1111/j.1469-8137.1967.tb05988.x.

Mullineaux P. M., Exposito-Rodriguez M., Laissue P. P., Smirnoff N. 2018. "ROS-dependent signalling pathways in plants and algae exposed to high light: Comparisons with other eukaryotes." *Free Radical Biology and Medicine*. 122: 52-64, https://doi.org/10.1016/j.freeradbiomed.2018.01.033.

Okamoto, K. O., Pinto, E., Latorre, L. R., Bechara, E. J. H., Colepicolo, P. 2001. "Antioxidant modulation in response to metal-induced oxidative stress in algal chloroplasts." *Archives of Environmental Contamination and Toxicology*. 40: 18–24.

Pilatti F. K., Ramlov F., Schmidt E. C., Costa Ch., a de Oliveira E. R., Bauer C. M., Rocha M., Bouzon Z. L., Maraschin M. 2017. "Metabolomics of *Ulva lactuca* Linnaeus (Chlorophyta) exposed to oil fuels: Fourier transform infrared spectroscopy and multivariate analysis as tools for metabolic fingerprint." *Marine Pollution Bulletin*. 114 2, 30: 831-836.

Pilatti F. K., F. Ramlov, E. C. Schmidt, M. K., D. T. Pereira, Ch. Costa, E. R. de Oliveira, Cl. M. Bauer, M. Rocha, Z. L. Bouzon, M. 2016. "*In vitro* exposure of *Ulva lactuca* Linnaeus (Chlorophyta) to gasoline–Biochemical and morphological alterations." *Chemosphere*. 156: 428-437. https://doi.org/10.1016/j.chemosphere.2016.04.126.

Pokora W., Tukaj Z. 2010. "The combined effect of anthracene and cadmium on photosynthetic activity of three Desmodesmus (Chlorophyta) species. *Ecotoxicology and Environmental Safety*. 73 6: 1207-1213. doi. org/10.1016/j.ecoenv.2010.06.013.

Pugovkin D. V., Liaimer A., Jensen J. B. 2016. "Epiphytic bacterial communities of the alga *Fucus vesiculosus* in oil-contaminated water areas of the Barents Sea." *Doklady Biological Sciences*. 471 1: 269-271.

Quadir P., J. Harrison, DeWreede R. E. 1979. "The effects of emergence and submergence on the photosynthesis and respiration of marine macrophytes." *Phycologia*. 18 1: 83-88. doi: 10.2216/i0031-8884-18-1-83.1.

Rogozhin V. V. 2004. *"Peroxidase as a component of the antioxidant system of living organisms."* St. Petersburg: GIORD. 240 p. [In Russia].

Rogozhin V. V., Verkhoturov, V. V., Kurilyuk, T. T. 2001. "The Antioxidant System of Wheat Seeds during Germination." *Biology Bulletin*. 28: 126–133. https://doi.org/10.1023/A:1009454713659.

Rossa, M. M., de Oliveira, M. C., Okamoto, O. K. 2002. "Effect of visible light on superoxide dismutase (SOD) activity in the red alga *Gracilariopsis tenuifrons* (Gracilariales, Rhodophyta)." *Journal of Applied Phycology*. 14:151–157. https://doi.org/10.1023/A:1019985722808.

Ryzhik I. V. 2016. "Seasonal Variations in the Metabolic Activity of Cells of *Fucus vesiculosus* Linnaeus, 1753 (Phaeophyta: Fucales) from the Barents Sea." *Russian Journal of Marine biology*. 42 5: 433-436. https://doi.org/10.1134/S1063074016050102.

Ryzhik I. V., Fisak, E. M. 2018. "Annual dynamics of the content of soluble phlorotannins in *Fucus vesiculosus* L. cells and their possible participation in tissue repair processes." *Questions of Modern Algology*. 1: 4-4.

Ryzhik I. V., Kosobryukhov A. A., Markovskaya E. F., Makarov M. V. 2021. "Photosynthetic Capacity of *Fucus vesiculosus* Linnaeus, 1753 (Phaeophyta: Fucales) in the Barents Sea during the Tidal Cycle." *Biol Bull Russ Acad Sci* 48:48–56. https://doi.org/10.1134/S1062359020060114.

Ryzhik I. V., Pugovkin D. V., Salakhov D. O., Klindukh M. P., Voskoboynikov G. M. 2022. "Physiological changes and rate of resistance of *Acrosiphonia arcta* (Dillwyn) Gain upon exposure to diesel fuel." *Heliyon.* 8 (8): e10177, https://doi.org/10.1016/j.heliyon.2022.e10177.

Schagerl M., Moostl M. 2011. "Drought stress, rain and recovery of the intertidal seaweed *Fucus spiralis*" *Mar Biol*. 158: 2471–2479. Doi: 10.1007/s00227-011-1748-x.

Schagerl M. 2011. "Drought stress, rain and recovery of the intertidal seaweed *Fucus spiralis*" *Mar. Biol.* (Berlin). 158: 2471–2479. https://doi.org/10.1007/s00227-011-1748-x.

Schmitz K., Lobban C. S. 1976. "A Survey of Translocation in Laminariales (Phaeophyceae)." *Marine Biology*. 36: 207-216.

Schmitz K., Srivastava L. M. 1979. "Long distance transport in *Macrocystis integrifolia* I. Translocation of 14C-labelled assimilates." *Plant Physiology*. 63 6: 995–1022.

Shakhmatova O. A., Milchakova N. A. 2009. "Catalase Activity of black sea species Cystoseira C. Ag. in various environmental conditions." *Algology*. 19 1: 34 – 46.

Shakhmatova O. A., Milchakova N. A., 2014. "Effect of environmental conditions on Black Sea macroalgae catalase activity." *Int. J. Algae*. 16 4: 377e391. http://doi.org/10.1615/InterJAlgae.v16.i4.70.

Stepanyan O. V. 2019. "Daily dynamics of photosynthesis, respiration and production of *Fucus vesiculosus* L. in the Barents Sea." *Algologia*. 29 4:440-445. https://doi.org/10.15407/alg29.04.440.

Thomas T. E., Turpin D. H., Harrison P. J. 1987. "Desiccation enhanced nitrogen uptake rates in intertidal seaweeds." *Marina Biology*. 94: 293–298. https://doi.org/10.1007/BF00392943.

Titlyanov E. A., Titlyanova T. V., Yakovleva I. M., Kalita T. L. 2006. "Rhythmical changes in the division and degradation of symbiotic algae in hermatypic corals." *Russian Journal of Marine Biology*. 32 1: 12-19. https://doi.org/10.1134/S1063074006010020.

Torres M. A., Barros M. P., Campos S. C. G., Pinto E., Rajamani S., Sayre, R. T., Colepicolo P. 2008. "Biochemical biomarkers in algae and marine pollution: a review." *Ecotoxicology and Environmental Safety*. 71 1: 1– 15.

Vardi A., Berman-Frank I., Rozenberg T., Hadas O., Kaplan A., Levine A. 1999. "Programmed cell death of the dinoflagellate (*Peridinium gatunense*) is mediated by CO_2 limitation and oxidative stress." *Current Biology*. 9 18:1061–1064.

Vega-López A., Griselda A. L., B. P. Posadas-Espadas, H. F. Olivares-Rubio, R. Dzul-Caamal. 2013. "Relations of oxidative stress in freshwater phytoplankton with heavy metals and polycyclic aromatic hydrocarbons." *Comparative Biochemistry and Physiology Part A: Molecular & Integrative Physiology*. 165 4: 498-507, https://doi.org/10.1016/j.cbpa.2013.01.026.

Waszczak C., Carmody M., & Kangasjärvi J. 2018. "Reactive oxygen species in plant signaling." *Annual review of plant biology*. 69: 209-236.

Wheeler P. A., North W. J. 1981. "Nitrogen supply, tissue composition and frond growth rates for *Macrocystis pyrifera* of the coast of southern California." *Marine Biology*. 64: 59-69.

Wiencke C., Gómez I., Dunton K. 2009. "Phenology and seasonal physiological performance of polar seaweeds." *Botanika Marina*. 52:6:585–592. https://doi.org/10.1515/BOT.2009.078.

Willekens H., Inzé D., Van Montagu M., Van Camp W. 1995. "Catalases in plants." *Molecular Breeding*. 1 3: 207-228.

Wolfe M. F., Schwartz G. J. B., Singaram S., Mielbrecht E. E., Tjeerdema R. S., Sowby M. L. 1998. "Influence of dispersants on the bioavailability of naphthalene from the water-accommodated fraction crude oil to the golden-brown algae, *Isochrysis galbana*." *Archives of environmental contamination and toxicology*. 35 2:274–280.

Biographical Sketch

Ryzhik Inna

Affiliation: Murmansk Marine Biological Institute of Russian Academy of Sciences

Education: Petrozavodsk State University

Business Address: Vladimirskaya St. 17 Murmansk 183010
Phone: +78152253963 e-mail: mmbi@mmbi.info

Research and Professional Experience:

Ecology and physiology of seaweed.

Stress physiology: the impact of abiotic and biotic environmental factors, such as temperature, salinity, oil pollution, etc., biotic relationships.

The antioxidant system of algae

Sanitary (bioremediation) and food aquaculture of seaweed, processing of seaweed.

Professional Appointments: Senior Researcher

Publications from the Last 3 Years:

1. Ryzhik I. Makarov M. Effect of diesel fuel film on green algae *Ulva lactuca* L. and *Ulvaria obscura* (Kützing) Gayral ex Bliding of the Barent Sea//*IOP Conf. Series: Earth and Environmental Science* / 4th International Scientific Conference "Arctic: History and Modernity" – 2019. – Vol. 302. – N 012029. – doi: 10.1088/1755-1315/302/1/012029.

2. Voskoboynikov G. M., Metelkova L. O., Makarov M. V., Ryzhik I. V., Pugovkin D. V. The algae-macrophytes of the Barents Sea in bioremediation of marine environment of oil product//*IOP Conf. Series: Earth and Environmental Science* (2019, April) (Vol. 263, No. 1, p. 012006). IOP Publishing. doi: 10.1088/1755-1315/263/1/012006.
3. Mityaev M. V., Gerasimova M. V., Ryzhik I. V., Ishkulova T. G. Insoluble fractions of aerosols and heavy metals in fresh snow in the North-West of the Kola Peninsula in 2018 *Ice and Snow*. 2019;59(3):307-318. (In Russ.) https://doi.org/10.15356/2076-6734-2019-3-386.
4. Ryzhik I. Voskobojnikov G. M. Pugovkin D., Pugovkin D., Makarov M., Michael Y. Roleda, L. Basova Tolerance of *Fucus vesiculosus* exposed to diesel water-accommodated fraction (WAF) and degradation of hydrocarbons by the associated bacteria//*Environmental Pollution*, 254. (2019) 113072 doi: 10.1016/j.envpol.2019.113072.
5. Voskoboinikov G. M., Ryzhik I. V., Salakhov D. O., Metelkova L. O., Zhakovskaya Z. A., Lopushanskaya E. M. Absorption and conversion of diesel fuel by the red alga *Palmaria palmata* (Linnaeus) F. Weber et D. Mohr, 1805 (Rhodophyta): the potential role of alga in bioremediation of sea water//*Russian Journal of Marine Biology*. 2020. V. 46. № 2. P. 113-118. https://doi.org/10.1134/S1063074020020108.
6. Voskoboinikov G. Salakhov D., Pugovkin D., Ryzhik I. The influence of diesel fuel on morpho-functional state of *Ulvaria obscura* (Chlorophyta)// *IOP Conference Series: Earth and Environmental Science*., 2020. – V. 539. – №. 1. – P. 012202. DOI: 10.1088/1755-1315/539/1/012202.
7. Shakhmatova O. Ryzhik I. Seasonal dynamics of catalase activity in *Cystoseira crinita* (Black Sea) and *Fucus vesiculosus* (Barents Sea) //*Ecological Chemistry and Engineering* (Chemia i Inzynieria Ekologiczna). – 2020. – Vol. 27, issue 4. – P. 643–650. – DOI: https://doi.org/10.2478/eces-2020-0041.
8. Ryzhik I. Dobychina E., Klindukh M., Machkarina O., Glukhikh Ya. *Seasonal Changes in the Concentration of Photosynthetic Pigments Palmaria palmata (Linnaeus) F. Weber & D. Mohr*//KnE Life Sciences: International applied research con-ference "Biological Resources Development and Environmental Management (BRDEM-2019)" (21 June 2019) / Ed. P. Makarevich. – 2020. – Vol. 2020. – P. 781–790. – Published 2020-01-15 – doi: 10.18502/kls.v5i1.6170.
9. Ryzhik I. Dobychina E., Klindukh M., Salahov D., Glukhikh Ya Usage of En-zymes of Algae-macrophytes Antioxidant System for Monitoring Water Pollution By Oil Products//KnE Life Sciences: International

applied research conference "*Bio-logical Resources Development and Environmental Management* (BRDEM-2019)" (21 June 2019) / Ed. P. Makarevich. – 2020. – Vol. 2020. – P. 808–818. – Published 2020-01-15 – doi: 10.18502/kls.v5i1.6176.

10. Ryzhik I. V., Makarov M. V., Kosobryukhov A. A., Markovskaya E. F.. Photosynthetic capacity of *Fucus vesiculosus* Linnaeus, 1753 (Phaeophyta: Fucales) in the Barents sea during the tidal cycle// *Biology Bulletin*. 2021. V. 48. № 1. P. 48-56. https://doi.org/10.1134/S10623590 20060114.
11. Ryzhik I. V, Klindukh M. P., Dobychina E. O. The B-group vitamins in the red alga *Palmaria palmata* (Barents Sea): Composition, seasonal changes and influence of abiotic factors//*Algal Research*. – 2021. –Vol. 59. – Article No 102473. – doi: 10.1016/j.algal.2021.102473.
12. Ryzhik I. V. Klindukh M., Dobychina E., Makarov M., Influence of diesel fuel on the composition and content of free amino acids in the green alga *Acrosiphonia arcta*//IOP Conf. Series: Earth and Environmental Science: AFE-2021 (Fundamental and Applied *Scientific Research in the Development of Agriculture in the Far East* (AFE 2021) 20-21 June 2021, Ussurijsk, Russian Federation). – 2021. – Vol. 937. – Article No 022010. – P. 1–5. – doi: 10.1088/1755-1315/937/2/022010.
13. Ryzhik I. V Salakhov D., Pugovkin D., Voskoboinikov G. The changes in the morpho-functional state of the green alga *Ulva intestinalis* L. in the Barents Sea under the influence of diesel fuel//*IOP Conf. Series: Earth and Environmental Science: AFE-2021 (Fundamental and Applied Scientific Research in the Development of Agriculture in the Far East (AFE 2021)* 20-21 June 2021, Ussurijsk, Russian Federation). – 2021. – Vol. 937. – Article No 022059. – P. 1–8. – doi: 10.1088/1755-1315/937/2/022059.
14. Ryzhik I. V M. Klindukh, E. Dobychina, M. Menshakova Physiological State of the Red Alga *Palmaria palmata* (L.) F. Weber et D. Mohr in the Barents Sea During the Winter Period//*International Journal on Algae*. Volume 23, 2021 Issue 4 doi: 10.1615/InterJAlgae.v23.i4.70.
15. Klindukh M., Ryzhik I., Makarov M. Changes in physiological and biochemical parameters of Barents sea *Fucus vesiculosus* Linnaeus 1753 in response to low salinity//*Aquatic Botany*. 2022. V. 176. P. 103469. https://doi.org/10.1016/j.aquabot.2021.103469.

Mikhail Makarov

Affiliation: Murmansk Marine Biological Institute of Russian Academy of Sciences

Education: Saint-Petersburg State University

Business Address: Vladimirskaya St.17 Murmansk 183010
Phone: +78152253963 e-mail: mmbi@mmbi.info

Research and Professional Experience:

Seaweed ecology and physiology, processes of phytocenoses damage and recruitment.

Reproductive biology of seaweed.

Stress physiology: impact of abiotic and biotic environmental factors, i.e., UVR, irradiation, temperature, salinity, hydrostatic pressure, oil pollution, etc.

Photobiology: photosynthesis, photosynthetic pigments.

Sanitary (bioremediation) and food aquaculture of seaweed, seaweed processing.

Professional Appointments: Research leader, Acting director

Publications from the Last 3 Years:

1. Voskoboynikov G. M., Metelkova L. O., Makarov M. V., Ryzhik I. V., Pugovkin D. V. The algae-macrophytes of the Barents Sea in bioremediation of marine environment of oil product//*IOP Conf. Series: Earth and Environmental Science* (2019, April) (Vol. 263, No. 1, p. 012006). IOP Publishing. doi: 10.1088/1755-1315/263/1/012006.
2. Ryzhik I. Makarov M. Effect of diesel fuel film on green algae *Ulva lactuca* L. and *Ulvaria obscura* (Kützing) Gayral ex Bliding of the Barent Sea//IOP Conf. Series: Earth and Environmental Science / 4th *International Scientific Conference "Arctic: History and Modernity"* – 2019. – Vol. 302. – N 012029. – doi: 10.1088/1755-1315/302/1/012029.
3. Ryzhik I. Pugovkin D., Makarov M., Michael Y. Roleda, L. Basova, Voskobojnikov GM Tolerance of *Fucus vesiculosus* exposed to diesel water-accommodated fraction (WAF) and degradation of hydrocarbons by the associated bacteria//*Environmental Pollution*, 254. (2019) 113072 doi: 10.1016/j.envpol.2019.113072.

4. Ryzhik I. V., Makarov M. V., Kosobryukhov A. A., Markovskaya E. F. Photosynthetic capacity of *Fucus vesiculosus* Linnaeus, 1753 (Phaeophyta: Fucales) in *the Barents sea during the tidal cycle//* Biology Bulletin. 2021. V. 48. No. 1. P. 48-56. https://doi.org/10.1134/S1062359020060114.
5. Ryzhik I. V., Klindukh M., Dobychina E., Makarov M., Influence of diesel fuel on the composition and content of free amino acids in the green alga *Acrosiphonia arcta//IOP Conf. Series: Earth and Environmental Science: AFE-2021* (Fundamental and Applied Scientific Research in the Development of Agriculture in the Far East (AFE 2021) 20-21 June 2021, Ussurijsk, Russian Federation). – 2021. – Vol. 937. – Article No 022010. – P. 1–5. – doi: 10.1088/1755-1315/937/2/022010.
6. Klindukh M., Ryzhik I., Makarov M. Changes in physiological and biochemical parameters of Barents Sea *Fucus vesiculosus* Linnaeus 1753 in response to low salinity//*Aquatic Botany*. 2022. V. 176. P. 103469. https://doi.org/10.1016/j.aquabot.2021.103469.

Chapter 3

Plant Catalases under Abiotic Stress: On Overview

Mouna Ghorbel[1,2] and Faiçal Brini[2,*]

[1]Biology Department, Faculty of Science, University of Hail, Hai'l, Saudi Arabia
[2]Laboratory of Biotechnology and Plant Improvement, Center of Biotechnology of Sfax, Sfax, Tunisia

Abstract

Hydrogen peroxide (H_2O_2) is produced in plants under normal and stressed conditions. At low concentration, this signal molecule is involved in different biological/physiological processes such as growth and development, cell cycle, photosynthetic functions, and plant responses to biotic and abiotic stresses. Interestingly, accumulation of H_2O_2 can cause Oxidative stress and then eventual cell death. Thus, plants developed a preferment antioxidant enzymatic system to defend against threats. Enzymatic antioxidant defense occurs as a series of redox reactions for ROS elimination. Different enzymes are implicated in this process such as catalase (CAT), ascorbate peroxidases (APX), and superoxide dismutase (SOD). Catalase is a crucial enzyme in antioxidant defense system protecting eukaryotes from oxidative stress. Those proteins are present in almost all living organisms and play important roles in controlling plant response to biotic and abiotic stresses by catalyzing the decomposition of H_2O_2. Based on recent reports, this chapter highlights the role of catalase in plant defense against accumulation of H_2O_2 to regulate plant response to different abiotic stresses.

[*] Corresponding Author's Email: faical.brini@cbs.rnrt.tn.

In: Catalase and Its Applications
Editor: Kaley Rutherford
ISBN: 979-8-88697-421-8

© 2022 Nova Science Publishers, Inc.

Keywords: abiotic stress, catalase, H_2O_2, oxidative stress, redox signaling, ROS

Introduction

Reactive oxygen species (ROS) are unpreventable byproducts of cell metabolism. ROSs has crucial signaling roles in living organisms in normal and unfavorable environmental conditions (Baxter et al., 2014; Waszczak et al., 2018). ROS are produced from atmospheric oxygen by its partial reduction, which occurs in the presence of electron donors. Plant ROS are generated by the reduction of monovalent atmospheric oxygen, in chloroplasts (Pospíšil, 2016; Foyer, 2018), in mitochondria (Gleason et al., 2011) and in peroxisomes during photorespiration (del Río et al., 2006; del Río and López-Huertas, 2016). In the apoplast, ROS are produced by plasma membrane-localized NADPH oxidase (Sagi and Fluhr, 2006), and oxalate oxidase (Voothuluru and Sharp, 2013). ROS could also be produced by the degradation of spermidine under the control of polyamine oxidase (Geilfus et al., 2015). ROS can be propagated from cell to cell. They transduce signals for long distances due to a process known as ROS waves (Mittler et al., 2011). Those waves are controlled by the interplays between oxidative stress induced Ca^{2+} fluxes, calcium (Ca^{2+}) channels, and NOX (Gilroy et al., 2016).

In tiny concentrations, ROS apply signaling functions such as regulating developmental processes (Xia et al., 2015; Mhamdi and Van Breusegem, 2018), controlling morphogenetic processes along with some phytohormones such as auxins, ABA, and cytokinins (Xia et al., 2015; Zwack et al., 2016). In another hand, auxin and ABA can regulate ROS production by activating NOX (Joo et al., 2001; Schopfer et al., 2002; Pasternak et al., 2007). Interestingly, ROS may affect auxin levels (Takác et al., 2016). In High concentrations, ROS may cause destructive oxidative effects on proteins, lipids, and nucleic acids which lead to cell death. Besides, ROS are implicated in the programmed cell death (Lehmann et al., 2015; Petrov et al., 2015). Those components lead to the expression of different stress-responsive genes. ROS have different types, but four ROS are more abundant and the most stable named as hydrogen peroxide (H_2O_2), singlet oxygen ($^1O_2$), hydroxyl radical ($OH\cdot$), and superoxide (O_2^-) (Dvorák et al., 2021). H_2O_2 is the most stable ROS. It is transported actively by aquaporins across membranes (Mittler, 2017; Smirnoff and Arnaud, 2019).

Cells contain different potent ROS scavengers such as glutathione peroxidases, peroxi-, thio-, and gluta-redoxins (Dietz, 2011; Kang et al., 2019; Foyer and Noctor, 2020). In another hand, the main antioxidant enzyme classes in cells are ascorbate peroxidases (APXs), catalases (CATs), dehydroascorbate reductases (DHARs), glutathione reductases (GRs), Superoxide dismutases (SODs), and monodehydroascorbate reductases (MDHARs) (Dietz, 2011; Kang et al., 2019; Foyer and Noctor, 2020; Dvorák et al., 2021).

Superoxide dismutases (SODs) are the major antioxidant enzymes. Those proteins could be cytosolic, mitochondrial, plastidic, or Peroxisomal, which are able to decompose O_2^- to H_2O_2 (Alscher et al., 2002; Pilon et al., 2011). In plants, SODs could be classified into three different classes based on the metal cofactors present in their active site: Cu/ZnSOD, MnSOD, FeSOD (Pilon et al., 2011). Based on the presence of n their active site, four different SODs are recognized in living organisms, namely NiSOD (not present in higher plants), and in the *Arabidopsis thaliana* genome, three FeSOD (FSD1, FSD2, and FSD3), one MnSOD (MSD1), and three Cu/ZnSOD (CSD1, CSD2, and CSD3; Kliebenstein et al., 1998). SODs are implicated in plants response to salt and oxidative stresses (Myouga et al., 2008; Xing et al., 2013; Shafi et al., 2015; Gallie and Chen, 2019; Dvorák et al., 2020). They are also implicated in seeds germination (Dvorák et al., 2020), lateral root growth (Morgan et al., 2008; Dvorák et al., 2020), and chloroplast development, and flowering (Rizhsky et al., 2003; Myouga et al., 2008).

Overproduced H_2O_2 is further converted into H_2O and O_2 by different classes of peroxidase (POD) and catalase (CAT) to water and O_2 (Tuzet et al., 2019). Catalases (CAT; EC 1.11.1.6) have a great affinity to H_2O_2. Those homotetrameric proteins are iron-containing enzymes implicated in plants responses to different developmental processes such as leaf development and senescence (Mhamdi et al., 2010; Zhang et al., 2020a), root growth (Yang et al., 2019), as well as pollen, ovule, shoot, and seed development (Sharma and Ahmad, 2014; Su et al., 2018; Palma et al., 2020).

Animal genomes contain a single catalase gene, whereas, in plants, this enzyme is encoded by a multigene family yielding in multiple isoenzymes. Using 12 genomic sequences from 6 plants, the comparison of the exon-intron structure of 12 genomic sequences was conducted. It was assumed that the ancestral gene coding for CAT had 7 introns (Iwamoto et al. 1998). Monocots-dicots evolutionary divergence, the occurrence of consecutive duplications of the primordial gene, and the differential loss of introns resulted in diverse isozyme genes (Iwamoto et al., 1998). In *Zea mays*, 3 genes were isolated

CAT1, CAT2, and CAT3. ZmCATs genes sequencing revealed the conservation of an identical coding region with variable introns (Abler and Scandalios, 1993; Guan and Scandalios, 1993; 1995), and the CAT1, CAT2, and CAT3 genes contained 6, 5, and 2 introns, respectively with an identical positioning of the introns suggesting an evolutionary link between all three *Zea mays* CAT genes (Guan and Scandalios, 1995; reviewed by Scandalios et al., 1997).

The number of catalase genes varies depending on the species, and their expression is regulated according to their tissue/organ distribution and the environmental conditions. So far, different catalase genes have been characterized in various plant species like *Gossypium hirsutum* L. and *Gossypium barbadense* L. (Wang et al., 2019), *Cucumis sativus* L. (Hu et al., 2016), rice (Iwamoto et al., 2000), *Triticum monococcum* (Tounsi et al., 2019), *Triticum turgidum* ssp durum (Garcia et al., 2000; Feki et al., 2015), *Triticum aestivum* (Zhang et al., 2022) and *Brassica napus* (Raza et al., 2021) (Table 1). For example, in *Arabidopsis* genome, three different genes encoding CATs (CAT1, CAT2, and CAT3) have been found. Interestingly, all three CAT isoforms are localized in peroxisomes (Frugoli et al., 1996; Du et al., 2008) and are crucial for the plant response to unfavorable conditions such as photo-oxidative stress (Zhang et al., 2020b). Moreover, CAT1, CAT2 and CAT3 generate a signal promoting autophagy-dependent cell death during plant immune responses (Hackenberg et al., 2013; Teh and Hofius, 2014). CAT1 isoform is crucial for plant response to the drought and salt stress (Xing et al., 2007), CAT2 is implicated in plant responses to heat; heavy metal; cold, and salt stresses (Bueso et al., 2007; Corpas and Barroso, 2017; Ono et al., 2020). CAT3 is also implicated in the drought stress response (Zou et al., 2015).

Recently, CRISPR/Cas9 method was used to generate at*cat1/3* double mutants and at*cat1/2/3* triple mutants. Analysis showed that the *cat1/2/3* triple mutants presented severe redox disturbance. Moreover, serious growth defects were also mentioned compared with wild-type plants and the at*cat2/3* double mutants. Additionally, the expression of different genes implicated in plants growth regulation as well as plants response to biotic and abiotic stresses were mentioned such as OXI*1* (*OXIDATIVE SIGNAL INDUCIBLE 1*) gene and several Mitogen Activated Protein Kinases cascade (Su et al., 2018). Interestingly, it has been shown that CAT2 present about 90% of the catalase activity of the proteins whereas *CAT1* and *CAT3* had a slight effect on catalase activity (Mhamdi et al., 2010). This indicates that CAT2 is the major catalase isoform that degrades H_2O_2 during photorespiration (Mhamdi et al., 2010).

Table 1. Number of catalase genes identified in different plant species

Species	Number of catalase genes	Roles	References
Gossypium hirsutum L.	7	-) *Verticillium dahlia* fungi infection -) Development and stress tolerance through modulating the reactive oxygen species (ROS) metabolism.	Wang et al., 2019
Gossypium. barbadense			
Triticum aestivum	10	-) High-regeneration trait of wheat immature embryos. -) Response to various stresses -) Antioxidant enzymes	She et al., 2013 Zhang et al., 2022
Arabidopsis thaliana	3	-) Reduced peroxisomal catalase activity increased sensitivity toward both ozone and photorespiratory H_2O_2-induced cell death in transgenic catalase-deficient. -) Drought stress conditions	Vandenabeele et al., 2004 Zou et al., 2015
Hordeum vulgare	2	-) Drought stress	Rohman et al., 2020
Zea mays	3	-) Pathogen contagion -) Oxidative stress	Wutipraditkul et al., 2011 Polidors et al., 2001
Capsicum annuum L.	3	-) Play a role in responses to environmental stresses.	Lee et al., 2005
Brassica napus L.	14	-) Respond to phytohormones, drought, low temperature, light, participate against defense.	Raza et al., 2021
Ipomea batatas L.	1	-) Plays a role in H_2O_2 homeostasis in leaves and in the plant's response to environmental stress	Chen et al., 2012
Nicotianaplum-baginifolia	3	-) Transgenic tobacco plants expressing the maize *CAT2* gene have enhanced the resistance to pathogen infection	Willekens et al., 1994 Polidoros et al., 2001
Cucumis sativus L	4	-) Heat, polyethylene glycol (osmotic stress), cold and NaCl (salinity stress)	Hu et al., 2016 Zhou et al., 2017
Pinus sylvestris L.	?	-) CAT is elaborated in embryogenesis and cell death procedures	Vuosku et al., 2015
Oryza sativa	4	-) Abiotic stress conditions – alkali, cold, dehydration, drought, heat, salinity, and submergence; and in response to various pathogen infection	Alam and Ghosh, 2018
Ricinus communis	1	?	Wang et al., 2019
S. lycopersicum	1	?	Wang et al., 2019
Prunus persica	2	-) Role for Cat1 in photorespiration and for Cat2 in stress responses	Bagnoli et al., 2004

CATs can be expressed in different molecular forms (or isozymes) encoded by different genes (Scandalios, 1968). Those CATs forms in plants are usually expressed in almost all tissues and at different developmental stages (reviewed by Scandalios et al., 1997). The expression of CAT genes is regulated both temporally and spatially and depends on developmental (Kwon and An, 2001) and environmental oxidative stimuli (Su et al. 2014). Moreover, it has been shown that CAT genes could be regulated depending on the circadian rhythm (Kabir and Wang, 2011). Besides, CAT expression could be differentially expressed in plants organs (Purev et al., 2010; Kabir and Wang, 2011; Su et al., 2014). In *Panax ginseng*, PgCAT1 gene expressing was higher in leaves and stems comparing with roots (Purev et al., 2010). In sugarcane, ScCAT1 expression was the highest in buds, followed by stem epidermis and stems pith, and was the least in leaves (Su et al., 2014). Dark and light conditions and their intensity can modulate CAT activity at molecular level (Azpilicueta et al., 2007). In tomato, SlCAT1 gene was high in the stems and flowers. SlCAT1 presented a higher expression level during the late light phase, whereas the expression of SlCAT2, expressed more in the leaves, was high during the early dark phase (Kabir and Wang, 2011). In *Brassica napus* genome, 10 BnCAT genes presented high expression levels in tissues (leaf, root, stem, and silique) (Raza et al., 2021).

In this chapter, we provide insight about ROS generation, their biochemistry and the signaling mechanism in plants to effectively cope with abiotic stress. We summarize strategies/plant disease resistance mechanisms, the antioxidant defense systems, and the crosstalk of ROS with the signaling molecules.

Localization and Biochemical Characterization of Catalase

In higher plants, CAT is principally localized in all differentiated peroxisomes such as peroxisomes of leaves, roots, cotyledons, and glyoxysomes and unspecialized peroxisomes (Su et al., 2014; Tounsi et al., 2019). Protein localization on peroxisome is controlled by a signal peptide called PTS located at the C-terminal part of the protein. In durum wheat, it has been shown that the deletion of the PTS1 domain suppress the localization of the protein and TdCAT1 was in the cytosol (Tounsi et al., 2019). Other studies showed that CATs could be in the mitochondria too (Scandalios 1990; Heazlewood et al., 2004; Shugaev et al., 2011). In Zea mays, CAT3 was found to be in mitochondria (Roupakias et al., 1980). Higher activity was detected for putative mitochondrial ZmCAT isoform compared to other *Zea mays* isoforms (Havir and McHale, 1989). AtCAT2 and AtCAT3 isolated from *Arabidopsis* are also located in the mitochondria (Heazlewood et al., 2004). CATs have a very high turnover rate as 26 million of H_2O_2 molecules can be changed by one CAT molecule within one minute (Anjum et al., 2016; Kaushal et al., 2018). Moreover, CAT proteins have a degradation constant of approximately 0.263 day^{-1} (Eising and Süselbeck, 1991); The reactions catalyzed by CAT proteins are very fast (Deisseroth and Dounce, 1970). When the concentration of H_2O_2 is low (<10−6 M), CAT works in peroxidic mode. At high concentrations of H_2O_2 (>10−6), CAT acts as both acceptor and donor of hydrogen molecules. CAT is inhibited by different compounds such as azide, cyanide, and hydroxylamine which confirm the presence of a heme prosthetic group in CATs. Moreover, the CAT proteins are inhibited by amino triazole and β-Mercaptoethanol. Those findings suggest that a thiol group, found in the active center of CAT, participate in the reactions controlled by CATs.

Catalase Regulation in Cells

In plants, the expression of CAT genes is tiny regulated depending on their important roles. For example, in *Arabidopsis* seedlings, application of H_2O_2 or ABA strongly induces CAT1 expression (Xing et al., 2008). ABI5, a basic leucine zipper transcription factor positively regulates CAT1 expression (via binding to CAT1 promoter) to enhance seed germination (Bi et al., 2017). Besides, AtGBF1 (G-Box Binding Factor 1) represses the expression of At*CAT2* gene, which result in ROS accumulation. This induces leaf senescence and triggers pathogen defense (Giri et al., 2017). Catalase in plants could be regulated depending on the circadian rhythm, the tissue, the surrounding environments (Azpilicueta et al., 2007; Su et al., 2014). Moreover, those proteins could be also modulated by different ligands. In fact, it has been recently shown that catalase protein could be activated in presence of different divalent cations such as Mn^{2+}, Fe^{2+}, Cu^{2+}, Zn^{2+}, and Mg^{2+} (Ghorbel et al., 2022). Moreover, catalase proteins could interact with calmodulins in plants in Calcium dependent manner as shown in potato (Afiyanti and Chen, 2014), *Arabidopsis* (Yang and Poovaiah, 2002), and sweet potato (Chen et al., 2012). Recently, Ghorbel et al. (2022) showed that durum wheat catalase 1 (TdCAT1) was able to interact with calmodulins in calcium independent manner. Interestingly, this interaction stimulates the catalytic activity of

TdCAT1 in CaM dose dependent manner, but the biological significance of this interaction is still not well understood (Figure 1) (Ghorbel et al., 2022).

It has been shown that Catalase proteins could interact with different proteins such as diphosphate kinase 1 (NDK1) (Fukamatsu et al., 2003), triple gene block protein 1 (TGBp1), and Arabidopsis salt overly sensitive 2 (SOS2), which is required for salt tolerance, can interact with CAT2 and CAT3 at an unknown subcellular location in response to salt stress and H_2O_2 metabolism. In Arabidopsis, it has been shown that CATs are able to interact with many other proteins. In fact, it has been shown that Salt overly sensitive 2 (AtSOS2), implicated in plants response to salt stress, are able to interact with CAT2 and CAT3 but the localization of this interaction is still not known (Verslues et al., 2007).

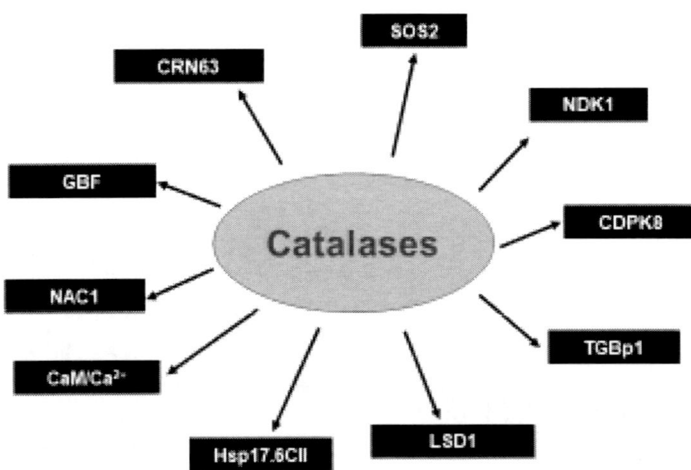

Figure 1. Schematic presentation of different proteins interacting Catalases in plants: the case of *Arabidopsis* and Tobacco. NDK1: diphosphate kinase 1; TGBp1: triple gene block protein 1; SOS2: salt overly sensitive 2; LSD1: zinc finger Lesion Simulating Disease1; Hsp14.6CII: small heat shock protein; NCA1: (NO Catalase Activity 1: a chaperone of catalase) dependent manner; CPK8: calcium dependent protein kinase 8; CRN63: Crinkling- And Necrosis-Inducing Protein 63; CaM/Ca^{2+}: Calmodulin Calcium complex.

CAT also interacts with the zinc finger Lesion Simulating Disease1 (LSD1) in Arabidopsis (Li et al., 2013). This interaction with all three CATs occurs in the cytoplasm and alters the activity of the proteins to modulate programmed cell death (Li et al., 2013). In *Arabidopsis*, it has been demonstrated that the small heat shock protein Hsp14.6CII interacts with catalase AtCAT2 in the cytosol (Li et al., 2015), as well as in the peroxisomes, and this interaction increases the catalytic activity of AtCAT2 in a NCA1 (NO Catalase Activity 1: a chaperone of catalase) dependent manner (Li et al., 2018). Moreover, NCA1 regulates CAT2 activity in response to cold, salt, and high pH via a RING-finger domain located at the N-terminal part of the protein and a C-terminal tetratricopeptide repeat-like helical domain (Li et al. 2015). Moreover, CAT2 interacts with a peroxisomal small heat shock protein and chaperone called Hsp17.6CII. This interaction occurs in peroxisome and positively regulates CAT2 activity in an NCA1-dependent manner (Li et al. 2017). Finally, phosphorylation of Arabidopsis CAT3 by CPK8 in response to drought stimulates the catalytic activity of AtCAT3 (Zou et al. 2015). In tobacco, NbCAT1 protein interacts with Crinkling- And Necrosis-Inducing Protein 63 (CRN63) with leads to NbCAT1 relocalization to nuclei. CRN63 is secreted by *Phytophthora sojae* under attack to regulate pathogen-induced cell death in tobacco (Zhang et al., 2015).

Catalase and Abiotic Stress Responses

Different stress factors, such as temperature, salinity, heavy metal, drought, etc., are known for generating an excess in ROS production and thus, disturb the equilibrium between ROS production and scavenging (Yamasaki et al., 2019). In *Brassica napus* genome, 4 genes BnCAT1–BnCAT3 and BnCAT11–BnCAT13 were up-regulated under other abiotic (cold, salinity), and hormonal (abscisic acid (ABA); gibberellic acid (GA)) treatments. Interestingly, the expression level of those catalase was not modified by drought and methyl jasmonate (MeJA) treatments. Interestingly, most of the 14 genes identified in brassica genome (14 genes) were up regulated by the waterlogging stress except three genes (BnCAT6, BnCAT9, and BnCAT10). (Raza et al., 2021).

Salt and Drought Stresses

Water deficit stress and salty soils are the major causes of limited plant maturation and productivity especially in semi-arid regions. Those stresses cause a complicated response at cellular, molecular, and physiological, such as antioxidant enzymatic (Table 2) (Anjum et al., 2016). Many studies indicated a positive correlation between catalase activity and the degree of stress intensity (Mhamdi et al., 2010). The catalase activity generally increases in presence of salt stress in the environment (Figure 2). For example, CAT activity increases proportionally with NaCl concentrations in white mulberry (*Morus alba* L.) subjected to salt stress (Sudhakar et al., 2001). This increase occurs in the acclimated *Morus alba* plants but not in the non-acclimated plants. Those founding suggest that acclimation may cause modifications in metabolic reactions taking place in peroxisomes. Moreover, it can be suggested that this may be related to the activation of processes preventing pro-oxidative changes under stressful conditions. The same result was also observed in cucumber plants (Naliwajski and Skłodowska, 2021), *Cucumis melo* var. Inodorus (Sivritepe et al., 2008). However, an increase in CAT in leaves peroxisomes were observed in a salt tolerant tomato variety was CAT activity increases after salt stress treatment (Mitova et al., 2002). In durum wheat (*Triticum turgidum* ssp durum) and *Triticum monococcum* subjected to salt stress, analysis showed that SOD and CAT proteins activities increased significantly with increasing concentrations of salt stress application (Tounsi et al., 2017). In another hand, it has been shown that CAT activity decreases in response to salt stress in many plants such as *Cicer arietinum* (Eyidogan and Oz, 2005), *Anabaena doliolum* (Srivastava et al., 2005), blackgram Vigna mungo (Sivakumar and Priya, 2019), and in a salt sensitive variety of rice *Oryza sativa* (Sharma et al., 2013).

Like salt stress, many studies revealed a positive correlation between CAT activity and the degree of drought applicated to plants (Mittler et al., 2011; Faize et al., 2011, Table 2). In *Panicum sumatrense* drought stress causes elongation of roots length but a reduction in plant height (Ajithkumar et al., 2014). Moreover, an important increase was registered in many endogen compounds such as free amino acid, glycine betaine and proline. Furthermore, this plant showed an increase in the antioxidant enzymes activities such as CAT to alleviate drought stress tolerance of this plant (Ajithkumar et al., 2014).

The same effect was also observed in *Jatropha curcas* were CAT activity, non-photochemical quenching system increase whereas leaf assimilation of CO_2 and lipid peroxidation dropped but the photochemical efficiency of PSII was unchanged in stressed conditions (Silva et al., 2019). Ford et al. (2011) studied the effect of drought stress on three different genotypes of Australian bread wheat (*Triticum aestivum* L.). All three cultivars presented an increase in ROS scavenging capacity and in oxidative stress metabolism.

Table 2. Effects of Salt stress and drought stress on catalase (CAT) activity in some plants

Plant species	Stress level	Stress-induced changes in CAT activity	References
Salt stress			
Triticum turgidum ssp durum	150 mM NaCl	Increase	Tounsi et al., 2017
Triticum monococcum	150 mM NaCl	Increase	
Cucumber (*Cucumis sativus* L.)	100 and 150 mM NaCl	Increase	Naliwajski and Skłodowska, 2021
bread wheat genotypes of Omani origin	40, 80 and 120 mM NaCl	Increase	AlHinai et al., 2022
Morus alba		Increase	Sudhakar et al., 2001
Oryza sativa	150 and 300 mM	Decrease	Sharma et al., 2013
Anabaena doliolum		Decrease	Srivastava et al., 2005
Vigna radiata		Decrease	Nahar et al., 2015
Mung beans (*Vigna radite*)	150 mM of NaCl	Increase	Ali et al., 2021
Solanum lycopersicum	0.04; 0.12 and 0.2 M	Increase	Sivakumar et al., 2020
Vigna mungo L.) variety TNAU (Blackgram) CO 6	125 mM NaCl	Decreases	Sivakumar and Priya, 2019
Jatropha curcas	1 M of NaCl, CaCl2 2 H2O, and MgCl2 6 H2O solution (Brazilian variety	Decrease	Compos et al., 2012
	100, 150, 200 mM NaCl (Chinese variety)	Increase	Gao et al., 2008
Salicornia europaea	550-1000 mM	Increase	Cárdenas-Pérez et al., 2022
Brassica napus	50, 100, 150, and 200 mM L-1 NaCl	Increase	El-Badri, et al., 2021
Cicer arietinum		Decrease	Eyidogan and Oz, 2005
Water deficit			
sugarcane		No change	Verma et al., 2019
Solanum lycopersicum ('Zhongza No.9')		Increase	Shi et al., 2016
Amaranthus tricolor		Increase	Sarker and Oba, 2018
Panicum sumatrense		Increase	Ajithkumar et al., 2014).
Strawberry (Fragaria × ananassa Duch.) cv. Kurdistan		Increase	Ghaderi et al. 2015
Cleome spinosa		Increase	Uzilday et al., 2012
Cleome gynandra		Increase	Uzilday et al., 2012
Jatropha curcas		Increase	Silva et al., 2019

In those varieties, CAT activities increased (Ford et al., 2011). *In Amaranthus tricolor*, application of drought stress caused a significant increase in CAT activity (Sarker and Oba, 2018). In *Solanum lycopersicum* ('Zhongza No.9') plants subjected to water deficit stress presented an increase of CAT activity after 3 days of stress application. This increase was not detected after 5 days of stress application (Shi et al., 2016). Moreover, Verma et al. (2019) reported that sugarcane plants subjected to drought stress had no significant change of CAT activities comparing with control plants. Interestingly, the CAT activity increased when Silicon was added to the medium.

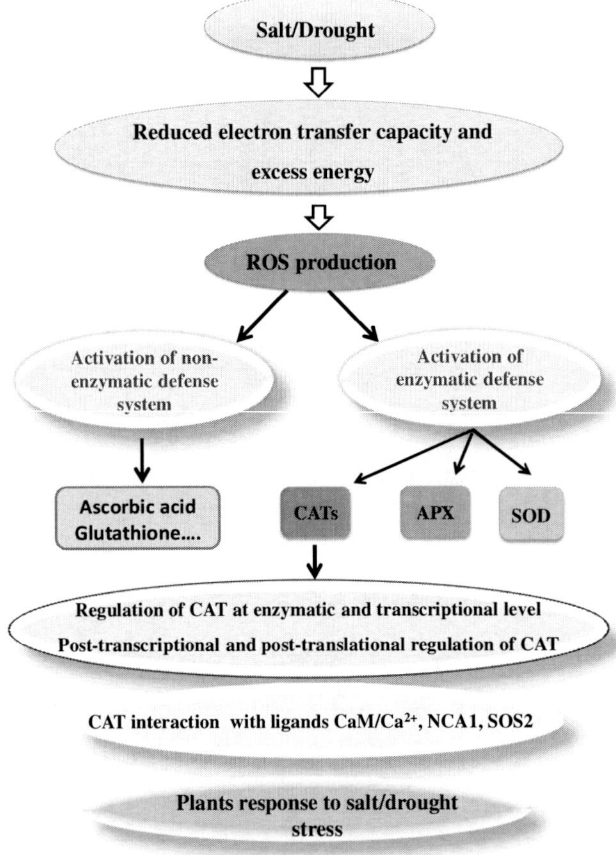

Figure 2. Schematic representation of the major biochemical and genetic effects of drought and saline stress in plants via the regulation of CAT activity.

Effect of Metals/Metalloids in CAT Activity

The effect of metal(oids) on CAT activities to alleviate metal-induced damage depends on plants species, metal type and/or concentration, duration of exposure, and plant age (the developmental stage) (Anjum et al., 2016). For example, application of cadmium stress on leaf disks of *Helianthus annuus* in light conditions enhanced the transcriptional level of the mitochondrial catalase CAT3. This effect was not shown when stress was applicated on etiolated plants (Azpilicueta et al., 2007). Several studies reported previously the modifications of CAT activities during metal stress suggesting different roles of CATs during plants responses to oxidative stresses caused by metals and metalloids (for more details see Anjum et al., 2016).

Cadmium Stress

In Cd stress conditions, *Brassica napus* plants showed an improved antioxidant system activity especially for CAT activities (Hasanuzzaman et al., 2012). The same effects were observed in *Brassica juncea* (Mobin and Khan, 2007), Coffea arabica (Gomes-Júnior et al. 2006), *Oryza sativa* (Hsu and Kao, 2004), and *Triticum aestivum* (Table 3) (Khan et al., 2007).

The same effects were also observed also in some macroalga including *Gracilaria tenuistipitata* (Collén et al., 2003), *Nannochloropsis oculata* (Lee and Shin, 2003), and *Ulva* fasciata (Wu et al., 2009). In some cases, Cd exposure causes a decrease in CAT activity. In soybean, exposed plants to such stress causes an increase in CAT activities in nodules and roots (Balestrasse et al., 2001). Moreover, an important accumulation level was observed under 100 μM $CdCl_2$ stress (Hossain et al., 2012). In *Zea mays* CAT activity was not modified after plants treatment with Cd (25 mM). Interestingly, plants pre-treatment with Salicylic acid (SA) for 6 hours causes a decrease in CAT activity to 50% (Krantev et al., 2008). The same effect was also observed when plants were treated with IAA (Agami and Mohamed, 2013). Interestingly, Cd-stressed *Pisum sativum* seedlings causes a significant increase in CAT activity at high Cd concentrations whereas at low Cd concentrations, CAT2 and CAT3, the most acidic isoforms, were enhanced by low Cd concentration (10–30 μM). The negative effect of high concentrations of Cd^{2+} were not recovered when Pisum plants were pre-treated with $CaCl_2$ mainly at high concentrations (5 mM) (El-Beltagi and Mohamed, 2013). In another hand, Cd stress (50 μg L^{-1}) induced the expression of CAT2 gene of the halophyte plant *Suaeda salsa* (Cong et al., 2013). *Sueda salsa* CAT2 genes was reported to play a crucial role in enhancing Catalase activity in plant under

Cadium stress as it is considered as a gene marker for Cd pollution (Cong et al., 2013).

Table 3. Effects of Heavy metal stress on catalase (CAT) activity in some plants

Plant species	Stress level	Stress-induced changes in CAT activity	References
Brassica rapa L. (Leaves of edible rape)	Foliar spray with 50 mg L−1 $CdCl_2$	Increase	Zong et al., 2017
Triticum aestivum L. cv. Pradip	0.5 mM Na_2HAsO_4 $.7H_2O$, 72 h	No change	Hasanuzzaman and Fujita, 2013
Raphanus sativus	Pb	Increase	El-Baltagi et al., 2010
Peanut	100 µM Arsenate (As)	Increase	Bianucci et al., 2017
Vigna radiata cv. Binamoog-1	1 mM$CdCl_2$, 48 h	Decreased by 55%	Hossain et al., 2010
Brassica juncea L. cv. Varuna	100 mg L−1, $CdSO_4{\cdot}8H_2O$), from 11 to 45 d	Decreased	Ahmad et al., 2011
Triticum aestivum L. cv. Giza168	500 µM Cd	Decrease by 20%	Agami and Mohamed, 2013
Brassica juncea Czern & Coss cv. T-59	150 µM $CdCl_2$	Increase	Hayat et al., 2007
Solanum nigrum	Cu	No change	Fidalgo et al., 2013
Kentucky bluegrass	50 µM $CdCl_2$, 7 d	Decrease by 11.2%	Guo et al., 2013
Helianthus annuus	100 µM $MnCl_2$, 4 d	Decreased by 45%	Saidi et al., 2014
	72 mM flurochloridone, 15 d	Decreased by 57%	Kaya and Yigi, 2014
Brassica napus cv. BINA sharisha 3	1mM $CdCl_2$, 48 h	Decreased by 28%	Hasanuzzaman et al., 2012
Prosopis farcta L	400 µM Pb	Increase	Zafari et al., 2017
poplar (*Populus deltoides* × *Populus nigra*)	Cd exposure	Increase	Zhang et al., 2014
Arabidopsis thaliana	Pb	Decrease	Corpas and Barroso, 2017
Zea mays L. cv. Norma	25 mM $CdCl_2$, 3 d	No change	Krantev et al., 2008

Other Metals/Metalloids

Arsenic stress also presented a positive effect in *Triticum aestivum* seedlings were CAT activities increased significantly (Hasanuzzaman and Fujita, 2013). This CAT activity enhancement was more pronounced in presence of Sodium nitroprusside, a NO donor (Hasanuzzaman and Fujita, 2013). *Arabidopsis thaliana* seedlings subjected to 100 μM $Pb(NO_3)_2$ presented an increase in NO content by 30% comparing with control plants (Corpas and Barroso, 2017). During this stress, catalase enzymes was inhibited in treated plants comparing with control. Thus, plomb stress effect varies depending on the plant as in *Prosopis farcta* L, Pb stress caused an increase in CAT activity (Table 3) (Zafari et al., 2017). An important increase in transcript level of tomato (*Solanum lycopersicum*) CATs SlCAT1 and SlCAT2 in Pb-exposed tomato (Kabir and Wang, 2011).

Manganese is a beneficial element for plants. However, the presence of elevated concentrations of Mn^{2+} (100 μM) negatively affects the activity of catalase proteins in *Helianthus annuus* as this activity decreased of about 45% (Saidi et al., 2014). More recently, it has been shown that the durum wheat catalase 1 gene (TdCAT1) was induced in plants under $MnCl_2$ stress (Feki et al., 2015). Some reports described that Mn^{2+} cations can increase the basal activity of CAT (Ghorbel et al., 2022) and upregulate CAT gene expression in many species (Tounsi et al., 2019; Zhou et al., 2013).

In Cu- and Zn-exposed *Vigna mungo*, a pronounced CAT activity was noted at germination stage, which dropped suggesting that CAT activity can be modified during the development stages of plants (Solanki and Poonam, 2011). PgCAT1, a CAT gene isolated from *Panax ginseng* was up regulated in plants under different heavy metals such as Cu (Purev et al., 2010). Furthermore, in *Prunus cerasifera*, the expression of CAT gene was stimulated under Cu stress (Lombardi and Sebastiani, 2005). In *Solanum nigrum* plants, the activity of CAT1 isoenzyme was more pronounced comparing with CAT2 in shoots. In the same plant, CAT2 activity was more important in roots comparing with CAT1 (Fidalgo et al., 2013). Cu stress inhibited the activity of both proteins and negatively affects the mRNA level of CAT2 in shoots but a greater accumulation in shoots (Fidalgo et al., 2013). ScCAT1, a catalase gene isolated from sugarcane genome positively regulated plants tolerance to Cu and Cd stress (Su et al., 2014). Application of Pb (20 μg/L) and/or Zn (100 μg/L) (alone or in combination) positively up-regulate the expression of CAT genes in *Suaeda salsa* (Wu et al., 2012).

Under Aluminum stress, the total activity of CAT was enhanced as well as the transcription of CAT genes in *Arabidopsis thaliana* (Richards et al.,

1998), and in *Capsicum annuum*, where the induction of CAT transcript was greater in the stem and during early stages of fruit development (Kwon and An, 2001).

Conclusion and Prospect

Catalase activity is generally modified in plants subjected to different abiotic stress. This activity could be stimulated, inhibited or mainly unchanged. Thus, it would be of great interest to consider that CAT activity in plants as a biochemical marker of plant tolerance to some stresses. Moreover, some reports described the interaction of different catalase proteins with other ligands (CaMs, NCA, cations….). It is important to discover the biological meaning of such relations to further understand plants responses to different abiotic stresses.

References

Afiyanti, M. and Chen, H. J. (2014). Catalase activity is modulated by calcium and calmodulin in detached mature leaves of sweet potato. *Journal of Plant Physiology,* 171, 35–47.

Agami, R. A. and Mohamed, G. F. (2013). Exogenous treatment with indole-3-acetic acid and salicylic acid alleviates cadmium toxicity in wheat seedlings. *Ecotoxicology and Environmental Safety*, 94, 164–171.

Ahmad, P., Nabi, G. and Ashraf, M. (2011). Cadmium-induced oxidative damage in mustard (*Brassica juncea* (L.) Czern. & Coss.) plants can be alleviated by salicylic acid. *South African Journal of Botany*, 77, 36–44.

Ajithkumar, I. P. and Panneerselvam, R. (2014). ROS scavenging system, osmotic maintenance, pigment, and growth status of *Panicum sumatrense* roth. under drought stress. *Cell Biochemistry and Biophysics.* 68, 587–595.

Al Hinai, M. S., Ullah, A., Al-Rajhi, R. S. and Farooq, M. (2022). Proline accumulation, ion homeostasis and antioxidant defence system alleviate salt stress and protect carbon assimilation in bread wheat genotypes of Omani origin. *Environmental and Experimental Botany,* 193, 104684.

Ali, R., Gul, H., Hamayun, M. Rauf, M., Iqbal, A., Shah, M., ussain, A., Bibi H. & In-Jung Lee. (2021). *Aspergillus awamori* ameliorates the physicochemical characteristics and mineral profile of mung bean under salt stress. *Chemical and Biological Technologies in Agriculture*, 8, 9.

Anjum, N. A., Sharma, P., Gill, S. S., Hasanuzzaman, M., Khan, E. A., Kachhap, K., Mohamed, A. A., Thangavel, P., Devi, G. D. and Vasudhevan, P. (2016). Catalase and ascorbate peroxidase—Representative H_2O_2-detoxifying heme enzymes in plants. *Environmental and Science Pollution Research,* 23, 19002–19029.

Azpilicueta, C. E., Benavides, M. P., Tomaro, M. L. and Gallego, S. M. (2007). Mechanism of CATA3 induction by cadmium in sunflower leaves. *Plant Physiology and Biochemistry*, 45, 589–595.

Balestrasse, K. B., Gardey, L., Gallego, S. M. and Tomaro, M. L. (2001). Response of antioxidant defence system in soybean nodules and roots subjected to cadmium stress. *Australian Journal of Plant Physiology*, 28, 497–504.

Baxter, A., Mittler, R. and Suzuki, N. (2014). ROS as key players in plant stress signalling. *Journal of Experimental Botany*, 65, 1229–1240.

Bi, C., Ma, Y., Wu, Z., Yu, Y. T., Liang, S., Lu, K. and Wang, X. F. (2017). *Arabidopsis* ABI5 plays a role in regulating ROS homeostasis by activating *CATALASE 1* transcription in seed germination. *Plant Molecular Biology*, 94, 197-213.

Bianucci, E., Furlan, A., del Carmen Tordable, M., Hernandez, L. E., Carpena-Ruiz, R. O. and Castro, S. (2017). Antioxidant responses of peanut roots exposed to realistic groundwater doses of arsenate: identification of glutathione S-transferase as a suitable biomarker for metalloid toxicity. *Chemosphere*, 181, 551–561.

Campos, M. L., de, O., Hsie, B. S., Granja, J. A. A., Correia, R. M. Almeida-Cortez, J. S. and Pompelli, M. F. (2012). Photosynthesis and antioxydant activity in *Jatropha curcas* L. under salt stress. *Brazilian Journal of Plant Physiology*, 24(1), 55-67.

Chen, H. J., Wu, S. D., Huang, G. J., Shen, C. Y., Afiyanti, M., Li, W. J. and Lin, Y. H. (2012). Expression of a cloned sweet potato catalase SPCAT1 alleviates ethephon-mediated leaf senescence and H_2O_2 elevation. *Journal of Plant Physiology,* 169, 86–97.

Collén, J., Pinto, E., Pedersén, M. and Colepicolo, P. (2003). Induction of oxidative stress in the red macroalga *Gracilaria tenuistipitata* by pollutant metals. *Archives of Environmental Contamination and Toxicology*, 45, 337–342.

Cong, M., Lv, J., Liu, X., Zhao, J. and Wu, H. (2013). *Gene expression responses in Suaeda salsa after cadmium exposure*. Springer Plus, 2, 232.

Corpas, F. J. and Barroso, J. B. (2017). Lead-induced stress, which triggers the production of nitric oxide (NO) and superoxide anion ($O2.^-$) in *Arabidopsis* peroxisomes, affects catalase activity. *Nitric Oxide*, 68, 103–110.

del Río, L. A. and López-Huertas, E. (2016). ROS Generation in peroxisomes and its role in cell signaling. *Plant Cell Physiology*, 57, 1364–1376.

del Río, L. A., Sandalio, L. M., Corpas, F. J., Palma, J. M. and Barroso, J. B. (2006). Reactive oxygen species and reactive nitrogen species in peroxisomes. Production, scavenging, and role in cell signaling. *Plant Physiology*, 141, 330–335.

Dvorák, P., Krasylenko, Y., Ovecka, M., Basheer, J., Zapletalová, V., Šamaj, J., & Takáč, T. (2020). *In-vivo* light-sheet microscopy resolves localisation patterns of FSD1, a superoxide dismutase with function in root development and osmoprotection. *Plant Cell Environment*, 44, 68–87.

Dvorák, P., Krasylenko, Y., Zeiner, A., Šamaj, J. and Takácˇ, T. (2021). Signaling Toward Reactive Oxygen Species-Scavenging Enzymes in Plants. *Frontiers in Plant Sciences*, 11, 618835.

El-Badri, A. M., Batool, M., A. A. Mohamed, I., Wang, Z., Khatab, A., Sherif, A., Ahmad, H., Khan, M. N., Hassan, H. M., Elrewainy, I. M., Kuai, J., Zhou, G., & Wang, B. (2021). Antioxidative and Metabolic Contribution to Salinity Stress Responses in Two Rapeseed Cultivars during the Early Seedling Stage. *Antioxidants*, 10, 1227.

El-Beltagi, H. S., Mohamed, A. A. and Ashed, M. M. R. (2010). Response of antioxidative enzymes to cadmium stress in leaves and roots of radish (*Raphanus sativus* L.). *Notulae Scientia Biologicae*, 2, 76–82.

El-Beltagi, H. S. and Mohamed, H. I. (2013). Alleviation of cadmium toxicity in *Pisumsativum* L. seedlings by calcium chloride. *Notulae Botanicae Horti Agrobotanici*, 41, 157–168.

Eyidogan, F. and Oz, M. T. (2005). Effect of salinity on antioxidant responses of chickpea seedlings. *Acta Physiologia Plantarum*, 29, 485–493.

Faize, M., Burgos, L., Faize, L., Piqueras, A., Nicolas, E., Barba-Espin, G., Clemente-Moreno, M. J., Alcobendas, R., Artlip, T. and Hernandez, J. A. (2011). Involvement of cytosolic ascorbate peroxidase and Cu/Zn-superoxide dismutase for improved tolerance against drought stress. *Journal of Experimental Botany,* 62, 2599–2613.

Feki, K., Kamoun, Y., Ben Mahmoud, R., Farhat-Khemakhem, A., Gargouri, A. and Brini, F. (2015). Multiple abiotic stress tolerance of the transformants yeast cells and the transgenic Arabidopsis plants expressing a novel durum wheat catalase. *Plant Physiology and Biochemistry*, 97, 420–431.

Fidalgo, F., Azenha, M., Silva, A. F., de Sousa, A., Santiago, A., Ferraz, P. and Teixeira, J. (2013). Copper-induced stress in *Solanum nigrum* L. and antioxidant defense system responses. *Food Energy Securety*, 2, 70–80.

Ford, K. L., Cassin, A. and Bacic, A. (2011). Quantitative proteomic analysis of wheat cultivars with differing drought stress tolerance. *Frontiers in Plant Science*, *2*, doi: 10.3389/fpls.2011.00044.

Foyer, C. H. (2018). Reactive oxygen species, oxidative signaling and the regulation of photosynthesis. *Environmental and Experimental Botany*, 154, 134–142.

Gallie, D. R. and Chen, Z. (2019). Chloroplast-localized iron superoxide dismutases FSD2 and FSD3 are functionally distinct in Arabidopsis. *PLoS ONE*, 14: e0220078.

Gao, S., Ouyang, C., Wang, S., Xu, Y., Tang, L. and Chen. F. (2008). Effects of salt stress on growth, antioxidant enzyme and phenylalanine ammonia-lyase activities in *Jatropha curcas* L. seedlings. *Plant and Soil Environment*, 54(9), 374–381.

Gilroy, S., Białasek, M., Suzuki, N., Górecka, M., Devireddy, A. R., Karpinski, S. and Mittler, R. (2016). ROS, Calcium, and Electric Signals: Key Mediators of Rapid Systemic Signaling in Plants. *Plant Physiology*, 171, 1606–1615.

Giri, M. K., Singh, N., Banday, Z. Z., Singh, V., Ram, H., Singh, D., Chattopadhyay, S. and Nandi, A. K. (2017). GBF1 differentially regulates *CAT2* and *PAD4* transcription to promote pathogen defense in *Arabidopsis thaliana. The Plant Journal*, 91, 802-815.

Ghaderi, N., Normohammadi, S. and Javadi, T. (2015). Morpho-physiological responses of strawberry (Fragaria × ananassa) to exogenous salicylic acid application under drought stress. *Journal of Agriculture Science and Technology*, 17, 167–178.

Ghorbel, M., Feki, K., Tounsi, S., Haddaji, N., Hanin, M. and Brini, F. (2022). The Activity of the Durum Wheat (*Triticum durum* L.) Catalase 1 (TdCAT1) Is Modulated by Calmodulin. *Antioxidants,* 11, 1483.

Gleason, C., Huang, S., Thatcher, L. F., Foley, R. C., Anderson, C. R., Carroll, A. J., Millar, A. H., & Singh, K. B. (2011). Mitochondrial complex II has a key role in mitochondrial-derived reactive oxygen species influence on plant stress gene regulation and defense. *Proceeding of National Academy of Sciences U.S.A.* 108, 10768–10773.

Gomes-Junior, R. A., Moldes, C. A., Delite, F. S., Pompeu, G. B., Gratão, P. L., Mazzafera, P., Lea, P. J., & Azevedo, R. A. (2006). Antioxidant metabolism of coffee cell suspension cultures in response to cadmium. *Chemosphere,* 65, 1330–1337.

Guan, L. and Scandalios, J. G. (1993). Characterization of the catalase antioxidant defense gene Cat1 of maize and its developmentally regulated expression in transgenic tobacco. *The Plant Journal*, 3, 527–536.

Guan, L. and Scandalios, J. G. (1995). Developmentally related response of maize catalase genes to salicylic acid. *Proceeding of National Academy of Sciences U.S.A.* 92, 5930–5934.

Guo, Q., Meng, L., Mao, P. C., Jia, Y. Q. and Shi, Y. J. (2013). Role of exogenous salicylic acid in alleviating cadmium induced toxicity in Kentucky bluegrass. *Biochemical Systematics and Ecology*, 50, 269–276.

Hasanuzzaman, M., Nahar, K., Alam, M. M. and Fujita. M. (2012). Exogenous nitric oxide alleviates high temperature induced oxidative stress in wheat (*Triticum aestivum* L.) seedlings by modulating the antioxidant defense and glyoxalase system. *Australian Journal of Crop Science*, 6, 1314–1323.

Hasanuzzaman, M. and Fujita, M. (2013). Exogenous sodium nitroprusside alleviates arsenic-induced oxidative stress in wheat (*Triticum aestivum* L.) seedlings by enhancing antioxidant defense and glyoxalase system. *Ecotoxicology*, 22:584–596.

Hayat, S., Ali, B., Hasan, S. A. and Ahmad, A. (2007). Brassinosteroid enhanced the level of antioxidants under cadmium stress in *Brassica juncea*. *Environmental and Experimental Botany*, 60, 33–41.

Hossain, M. A., Hasanuzzaman, M. and Fujita, M. (2010). Up-regulation of antioxidant and glyoxalase systems by exogenous glycinebetaine and proline in mung bean confer tolerance to cadmium stress. *Physiology and Molecular Biology of Plants*, 16, 259–272.

Hsu, Y. T. and Kao, C. H. (2004). Cadmium toxicity is reduced by nitric oxide in rice leaves. *Plant Growth Regulation*, 42, 227–238.

Kaushal, J., Mehandia, S., Singh, G., Raina, A. and Arya, S. K. (2018). Catalase enzyme: Application in bioremediation and food industry. *Biocatalysis and Agricultural Biotechnology*, 16, 192–199.

Kaya, A. and Yigi, E. (2014). The physiological and biochemical effects of salicylic acid on sunflowers (*Helianthus annuus*) exposed to flurochloridone. *Ecotoxicology and Environmental Safety*, 106, 232–238.

Khan, N. A., Samiullah, S. S. and Nazar, R. (2007). Activities of antioxidative enzymes, sulphur assimilation, photosynthetic activity and growth of wheat (*Triticum aestivum*) cultivars differing in yield potential under cadmium stress. *Journal of Agronomy and Crop Science*, 193, 435–444.

Krantev, A., Yordanova, R., Janda, T., Szalai, G. and Popova, L. (2008). Treatment with salicylic acid decreases the effect of cadmium on photosynthesis in maize plants. *Journal of Plant Physiology*, 165, 920–931.

Lehmann, S., Serrano, M., L'Haridon, F., Tjamos, S. E. and Metraux, J. P. (2015). Reactive oxygen species and plant resistance to fungal pathogens. *Phytochemistry*, 112, 54–62.

Li, J., Liu, J., Wang, G., Cha, J. Y., Li, G., Chen, S., Li, Z., Guo, J., Zhang, C., Yang, Y., Kim, WY., Yun, DJ., Schumaker, K. S., Zhongzhou Chen, and Guoa, Y.(2015). A Chaperone Function of NO CATALASE ACTIVITY1 Is Required to Maintain Catalase Activity and for Multiple Stress Responses in Arabidopsis. *Plant Cell*, 27, 908–925.

Li, G., Li, J., Hao, R. and Guo, Y. (2017). Activation of catalase activity by a peroxisome-localized small heat shock protein Hsp17.6CII. *Journal of Genetics and Genomics*, 44, 395–404.

Li, Y., Chen, L., Mu, J. and Zuo, J. (2013). LESION SIMULATING DISEASE 1 interacts with catalases to regulate hypersensitive cell death in *Arabidopsis*. *Plant Physiology*, 163, 1059-1070.

Li, G., Li, J., Hao, R. and Guo, Y. (2017). Activation of catalase activity by a peroxisome-localized small heat shock protein Hsp17.6C II. *Journal of Genetics and Genomics*, 44, 395-404.

Mhamdi, A., Queval, G., Chaouch, S., Vanderauwera, S., Van Breusegam, F. and Noctor, G. (2010). Catalase function in plants: A focus on *Arabidopsis* mutants as stress-mimic models. *Journal of Experimental Botany*, 61, 4197-4220.

Mittler, R., Vanderauwera, S., Suzuki, N., Miller, G., Tognetti, V.B., Vandepoele, K., Gollery, M., Shulaev, V. and Van Breusegem, F. (2011). ROS signaling: The new wave? *Trends in Plant Sciences,* 16, 300–309.

Mobin, M. and Khan, N. A. (2007). Photosynthetic activity, pigment composition and antioxidative response of two mustard (*Brassica juncea*) cultivars differing in photosynthetic capacity subjected to cadmium stress. *Journal of Plant Physiology*, 164, 601–610.

Myouga, F., Hosoda, C., Umezawa, T., Iizumi, H., Kuromori, T., Motohashi, R., Shono, Y., Nagata, N., Ikeuchi, M., & Shinozaki, K. (2008). A heterocomplex of iron superoxide dismutases defends chloroplast nucleoids against oxidative stress and is essential for chloroplast development in *Arabidopsis*. *The Plant Cell,* 20, 3148–3162.

Nahar, K., Hasanuzzaman, M., Alam, M. M. and Fujita, M. (2015). Regulatory roles of exogenous glutathione in conferring salt tolerance in mung bean (*Vigna radiata* L.): implication of antioxidant defense and methylglyoxal detoxification systems. *Biologia Plantarum*, 59, 745-756.

Petrov, V., Hille, J., Mueller-Roeber, B. and Gechev, T. S. (2015). ROS-mediated abiotic stress-induced programmed cell death in plants. *Frontiers in Plant Sciences*, 6, 69.

Pospíšil, P. (2016). Production of reactive oxygen species by photosystem II as a response to light and temperature stress. *Frontiers in Plant Sciences,* 7, 1950.

Raza, A., Su, W., Gao, A., Mehmood, S. S., Hussain, M. A., Nie, W., Lv, Y., Zou, X. and Zhang, X. (2021). Catalase (CAT) Gene Family in Rapeseed (*Brassica napus* L.): Genome-Wide Analysis, Identification, and Expression Pattern in Response to

Multiple Hormones and Abiotic Stress Conditions. *The International Journal of Molecular Sciences,* 22, 4281.

Saidi, I., Nawel, N. and Djebali, W. (2014). Role of selenium in preventing manganese toxicity in sunflower (*Helianthus annuus*) seedling. *South African Journal of Botany*, 94, 88–94.

Sarker, U. and Oba, S. (2018). Catalase, superoxide dismutase and ascorbate-glutathione cycle enzymes confer drought tolerance of *Amaranthus tricolor. Scientific Reports*, 8, 16496.

Sharma, I., Ching, E., Saini, S., Bhardwaj, R. and Pati, P. K. (2013). Exogenous application of brassinosteroid offers tolerance to salinity by altering stress responses in rice variety *Pusa basmati-1. Plant Physiology and Biochemistry*, 69, 17–26.

Shi, Y., Zhang, Y., Han, W., Feng, R., Hu, Y., Guo, J. and Gong, H. (2016). Silicon Enhances Water Stress Tolerance by Improving Root Hydraulic Conductance in *Solanum lycopersicum* L. *Frontiers in Plant Sciences*, 7, 196.

Silva, E. N., Silveira, J. A. G., Aragão, R. M., Vieira, C. F., & Carvalho, F. E. L. (2019). Photosynthesis impairment and oxidative stress in *Jatropha curcas* exposed to drought are partially dependent on decreased catalase activity. *Acta Physiologia Plantarum,* 41, 4.

Sivakumar, R. and Jaya Priya, S. (2019). PGRs and nutrient consortium effect on water relations, photosynthesis, catalase enzyme and yield of blackgram under salinity stress. *Legume Research*, 413-418.

Sivakumar, J., Prashanth, J. E. P., Rajesh, N., Reddy, S. M. and Pinjari, O. B. (2020). Principal component analysis approach for comprehensive screening of salt stress-tolerant tomato germplasm at the seedling stage. *Journal of Bioscience* 45, 141.

Smirnoff, N. and Arnaud, D. (2019). Hydrogen peroxide metabolism and functions in plants. *New Phytologist*, 221, 1197–1214.

Solanki, R. and Poonam, A. (2011). Zinc and copper induced changes in physiological characteristics of *Vigna mungo* (L.). *Journal of Environmental Biology*, 32, 747–751.

Srivastava, A. K., Bhargava, P. and Rai, L. C. (2005). Salinity and copper-induced oxidative damage and changes in antioxidative defense system of *Anabaena doliolum. World Journal of Microbiology and Biotechnology*, 22, 1291–1298.

Sudhakar, C., Lakshmi, A. and Giridarakumar, S. (2001). Changes in antioxidant enzyme efficacy in two high yielding genotypes of mulberry (*Morus alba* L.) under NaCl salinity. *Plant Science*, 161, 613–619.

Tounsi, S., Kamoun, Y., Feki, K., Jemli, S., Saïdi, M. N., Ziadi, H., Alcon, C. and Brini, F. (2019). Localization and expression analysis of a novel catalase from *Triticum monococcum* TmCAT1 involved in response to different environmental stresses. *Plant Physiology and Biochemistry*, 139, 366–378.

Uzilday, B., Turkan, I., Sekmen, A. H., Ozgur, R. and Karakaya, H. C. (2012). Comparison of ROS formation and antioxidant enzymes in *Cleome gynandra* (C4) and *Cleome spinosa* (C3) under drought stress. *Plant Science,* 182, 59–70.

Verslues, P. E., Batelli, G., Grillo, S., Agius, F., Kim, Y. S., Zhu, J., Agarwal, M., Katiyar-Agarwal S., and Zhu, JK. (2007). Interaction of SOS2 with nucleoside diphosphate kinase 2 and catalases reveals a point of connection between salt stress and H_2O_2 signaling in *Arabidopsis thaliana*. *Molecular Cell Biology,* 27, 7771-7780.

Waszczak, C., Carmody, M. and Kangasjärvi, J. (2018). Reactive oxygen species in plant signaling. *Annual Review of Plant Biology*, 69, 209–236.

Wu, H., Liu, X., Zhao, J. and Yu, J. (2012). Toxicological responses in halophyte Suaeda salsa to mercury under environmentally relevant salinity. *Ecotoxicology and Environmental Safety*, 85, 64–71.

Xing, Y., Jia, W. and Zhang, J. (2008). AtMKK1 mediates ABA-induced *CAT1* expression and H_2O_2 production via AtMPK6-coupled signaling in Arabidopsis. *The Plant Journal*, 54, 440-451.

Yamasaki, H., Ogura, M. P., Kingjoe, K. A. and Cohen, M. F. (2019). d-cysteine-induced rapid root abscission in the water fern Azolla Pinnata: Implications for the linkage between d-amino acid and reactive sulfur species (RSS) in plant environmental responses. *Antioxidants, 8*, 411.

Yang, T. and Poovaiah, B. W. (2002). Hydrogen peroxide homeostasis: Activation of plant catalase by calcium/calmodulin. *Proceeding of National Academy of Sciences U.S.A*, 99, 4097–4102.

Zafari, S., Sharifi, M., Mur, L. A. J. and Chashmi, N. A. (2017). Favouring NO over H_2O_2 production will increase Pb tolerance in *Prosopis farcta* via altered primary metabolism. *Ecotoxicology and Environmental Safety*, 142, 293–302.

Zhang, F., Wan, X., Zheng, Y., Sun, L., Chen, Q., Zhu, X., Guo, Y. L. and Liu, M. (2014). Effects of nitrogen on the activity of antioxidant enzymes and gene expression in leaves of Populus plants subjected to cadmium stress. *Journal of Plant Interaction*, 9, 599–609.

Zhang, M., Li, Q., Liu, T. et al. (2015). Two cytoplasmic effectors of *Phytophthora sojae* regulate plant cell death via interactions with plant catalases. *Plant Physiology*, *167*(1), 164-175.

Zhang, Y., Ji, T. T., Li, T. T., Tian, Y. Y., Wang, L. F. and Liu, W. C. (2020a). Jasmonic acid promotes leaf senescence through MYC2-mediated repression of CATALASE2 expression in Arabidopsis. *Plant Science*, *299*, 110604.

Zhang, S., Li, C., Ren, H. et al.. (2020b). BAK1 mediates light intensity to phosphorylate and activate catalases to regulate plant growth and development. *International Journal of Molecular Sciences*, *21*(4), 1437.

Zhang, M., Ji, P., Li, Z., Sun, Z., Tran, N. T. and Li, S. (2022). Catalase regulates the homeostasis of hemolymph microbiota and autophagy of the hemocytes in mud crab (Scylla paramamosain). *Aquaculture Reports*, *25*, 101237.

Zong, H. Y., Liu, S., Xing, R., Chen, X. and Li, P. C. (2017). Protective effect of chitosan on photosynthesis and antioxidative defense system in edible rape (*Brassica rapa* L.) in the presence of cadmium. *Ecotoxicology and Environmental Safety*, 138, 271–278.

Zou, J. J., Li, X. D., Ratnasekera, D., Wang, C., Liu, W. X., Song, L. F., Zhang, W. Z. and Wu, W. H. (2015). Arabidopsis CALCIUM-DEPENDANT PROTEIN KINASE 8 and CATALASE 3 function in abscisic acid-mediated signaling and H_2O_2 homeostasis in stomatal guard cells under drought stress. *The Plant Cell*, 27, 1445-1460.

Chapter 4

The Blood Catalase in the Human Body

Teréz Nagy*, PhD and László Góth, PhD
Department of Medical Imaging, University of Debrecen, Debrecen, Hungary

Abstract

The catalase enzyme is one of the earliest known and well studied enzymes. In the decades following its discovery in 1819, the enzyme catalase has been studied and detected in many types of living organisms, organs and tissues. Its substrate hydrogen peroxide was discovered in 1818. In the body, the catalase enzyme is responsible for eliminating hydrogen peroxide in toxic concentrations. The enzyme is encoded by a single gene (33114 Kb, 13 exons, 12 introns). Several polymorphisms and mutations have been identified for the catalase gene. Red blood cells contain 99.9% of blood catalase.

The activity of the blood catalase can be easily determined by a widely used spectrophotometric method. It is a combination of optimized enzymatic conditions and the spectrophotometric assay of hydrogen peroxide based on formation of its stable complex with ammonium molybdate which could be measured at 405 nm.

Acatalasemia occurs when the activity of the enzyme blood catalase is less than 10%. Hypocatalasemia is when blood cell activity is below 50% of the average. We have reported patients with acatalasemia and hypocatalasemia in diseases such as diabetes mellitus, microcyter anemia, vitiligo, and thalassemia. We've identified mutations and polymorphisms in the catalase gene that may be responsible for the decrease in enzyme activity. Mutations and polymorphisms are in exons and introns. The effects of the polymorphisms in the introns are currently

* Corresponding Author's Email: nyestene@.unideb.hu.

In: Catalase and Its Applications
Editor: Kaley Rutherford
ISBN: 979-8-88697-421-8
© 2022 Nova Science Publishers, Inc.

unknown. According to the exon mutations of catalase gene eleven types of acatalasemia were established in Hungary.

Enzyme catalase plays an important role in defense against the toxic effects of free oxygen species by the removal of hydrogen peroxide. Hydrogen peroxide may be formed in several pathologic conditions such as inflammation, cancer, erythrocyte metabolism, homeostasis and platelet activation.

In this chapter we are discussing: the catalase enzyme (discovery, history, the relationship with hydrogen peroxide and reactive oxygen species, the blood and erythrocyte catalase), the catalase gene (its deficiency acatalasemia and hypocatalasemia, its mutations), the blood catalase in various diseases (diabetes mellitus, anemia, thalassemia, vitiligo) and the possible future of catalase research.

Keywords: catalase enzyme, catalase gene, polymorphisms, mutations, acatalasemia

Catalase

In 1818 Gay-Lussac and Louis Jacques Thénard chemists observed that oxygen forms in the course of the decomposition of the hydrogen peroxide. Thénard observed, that plant vegetal and animal tissue are able to decompose the hydrogen peroxide. Schönbein detected in blood in 1863, Bergergrün identified for red blood cells in 1888, Gottstein and Beijerinck for bacteria in 1893 [1].

Beijerinck suggested the catalase test to separate bacteria into catalase negative and catalase positive strains according to their ability to decompose hydrogen peroxide. Nowadays, this test has been for clinical bacteriology.

The "Catalase" name appeared firstly in 1901 and it was proposed by Oscar Loew a German agricultural chemist working in USA. He wrote "Catalase a new enzyme of general occurrence." This name is associated with the catalytic ability of this enzyme due to the very fast reaction when liberated oxygen bubbles are coming from hydrogen peroxide [2].

The first measurements for determination of catalase activity were the titrimetric ones. They could detect catalase in human blood too and this test has been used since 1930. These methods were cheap and very simple because the substrate hydrogen peroxide and the product of oxygen could be measured by very simple methods.

The catalase enzyme (EC 1.11.1.6, hydrogen-peroxide: hydrogen-peroxide oxidoreductase) is made up of 4 symmetrical subunits with a tetramer structure and a size of 240 kDa. The subunits consist of 527 amino acids, containing a group of one-one hem, there are an iron (III) ions in the active centre. The three-channel structure ensures the optimal functioning of the enzyme. The main channel forms the way for the hydrogen peroxide substrate from the surface of the enzyme to the active centre. The narrow shape of the channel ensures that only hydrogen-peroxide and water molecules can pass and the others cannot. Oxygen molecules are released through the central channel. The determination of activity of enzyme catalase which decomposes hydrogen peroxide into oxygen and water. Several assays measure the decrease of its hydrogen peroxide substrate either directly at 240 nm or in coupled reactions with color-producing substances. The frequently used catalase assays are listed in Table 1. The first three methods are measuring the hydrogen peroxide substrate directly at 240 nm [3-5]. Their very frequent use is in case of pure catalase samples but is not preferred for samples with high proteins, turbidity, and colored ones. The other tests were the colorimetric ones [6-9]. Our method [7] was developed for human blood serum and its modification for human blood [8] with special interest their human reference ranges. This method gained the highest citation in 2020 and 2021. The fields of science which reported on this assay were Biochemistry-Genetics-Molecular Biology-Free Radicals followed by Medicine-Agriculture-Botany-Plants-Environmental Protection- Pharmacology-Toxicology. In July of, 2022 this assay was cited by 2016 times according to Google Scholar and 1729 times by Research Gate where its reads were 11 511.

Table 1. Catalase assays and their citations

Authors	Citation	Journal/Book chapter*	Year	Method	Reference
Bergmeyer HU	25.233	*Anal Biocmem*	1955	UV photomerty of H_2O_2	3
Aebi H	17.828	Methods in Enzymolgy*	1984	UV photomerty of H_2O_2	4
Beers S, Seizer IW.	7.149	*J Biol Chem*	1952	UV photomerty of H_2O_2	5
Sinha AK	5.790	*Anal Biochem*	1988	H_2O_2+$K_2Cr_2O_7$	6
Goth L.	2.015	*Clinica Chimica Acta*	1991	H_2O_2+ $(NH_4)6Mo_7O_{24}$	7
Johansson LH.	1.034	*Anal Biochem*	1988	CH-OH→CHO + Purpald	8

Hydrogen Peroxide and ROS

Hydrogen peroxide with its high concentration is toxic for DNA RNA, proteins, lipids, and tissues. It may be related to the formation of reactive oxygen species (ROS) and free radicals. Contrary, the new findings suggest that ROS and hydrogen peroxide to are important in some physiological processes such as cell circle regulation, activating and inhibiting certain steps of apoptosis and necrosis, and activation of enzymes [10].

High level of ROS could cause oxidative stress when ROS production exceeds the antioxidant capacity. The main sources of ROS are the mitochondrial electron transport chain, phagocyte cells and enzymes. The enzymes are nicotinamide-adenine-dinucleotide-phosphate(NADP) oxidase, cytochrome P450-oxidase from peroxisome, monoamine-oxidase, and xanthine-oxidase. The most commonly occurring ROS is hydroxyl radical (OH), nitrogen monoxide radical (NO), superoxide anion (O2-), lipid peroxide radical (LOO), and hydrogen peroxide (H_2O_2).

The human body developed defence mechanisms to reduce the damage of ROS. This system has low molecular weight antioxidants such as ascorbic acid, glutathione (GSH), alpha-tocopherol, and the antioxidant enzymes superoxide dismutase, glutathione peroxidase, and catalase to eliminate the free radicals. The third line of defence works at the molecular level when toxicity has already developed. It happens to remove the damaged section or correct it by repair enzymes, lipases, peptidases, proteases, and DNA repair enzymes.

Hydrogen peroxide is a small molecule that slightly exceeds the size of the water molecule has no charge and diffuses in the cell and between cells. It can be generated as a result of both physiological and pathological processes. Hydrogen peroxide concerns a "paradoxal effect" it is toxic for the body in high concentrations while its low concentration has important role in several physiological processes such as the activation of NFκB, signalling, carbohydrate metabolism, immune response, activation of protein kinase C and regulation of transcription factors.

By activating phagocytes in inflammatory tissues, hydrogen peroxide can be generated with a role of modulation the inflammatory process and promoting the expression of adhesion molecules. It can help the cell to proliferate, reach an apoptotic state, and modulate the aggregation of platelets.

Hydrogen peroxide is produced by xanthine oxidase and aldehyde oxidase in the cytosol, by the cytochrome P450 of autooxidation in the endoplasmic reticulum, uric acid oxidase, D-amino acid oxidase, hydroxyl-acid-oxidase in peroxisomes, monoamine oxidase and pyroxamine oxidase in the mitochondria. It may be produced by certain microbes such as lactic acid bacteria, as well as oxidase enzymes in urine. Hydrogen peroxide can be detected during mitochondrial electron transfer in apoptosis with contribution of the xanthin oxidase.

The main sources of hydrogen peroxide could be attributed to the superoxide dismutase that converts the highly toxic superoxide (O^-_2) anion into the less toxic H_2O_2. Superoxide anion is generated in ischemia, reperfusion, muscle loss, immobilization, and intense physical exertion. It is also formed in mitochondrial oxidative stress, malignant processes, diabetes, and other oxidative conditions.

For cells, hydrogen peroxide concentrations of 20 to 50 μmol/l are moderately toxic, while above 50 μmol/l are highly toxic.

To date, it is known that catalase is the main regulator of the hydrogen peroxide metabolism followed by the peroxidative effects of glutathione peroxidase, peroxiredoxin 2, and hemoglobin. They play major roles in removing H_2O_2 at relatively low concentrations, whereas the contribution of catalase increases when intracellular H_2O_2 is high [10-15].

Blood and Erythrocyte Catalase

The enzyme catalase is found in human organs at different concentrations. Its specific activity in human tissues are liver (peroxisomes, mitochondria) approximately 3.33x105 U/g, red blood cells about 4.67 x 105 U/g, heart has a medium concentration and the pancreas, the brain 0.78 and 0.722 x 105 U/g while its lowest concentration yields the serum with 56.7 kU/L. It may be supposed that tissues such as the liver and blood with their high catalase contents could provide antioxidant protection for the catalase-poor tissues [16].

In diseases of liver and erythrocytes, their high catalase content from their damaged tissues could get into the circulation increasing the serum catalase activity. Therefore, the increased serum catalase activity could be used in the diagnosis of their diseases.

The catalase content of the other blood cells is low when compared to that of erythrocytes [17, 18].

About 99% of blood catalase comes from red blood cells, and the average blood catalase is 113.3 MU per 1L of blood. It could be referred to the red blood cells it yields 251.7 MU/L. Therefore, blood catalase activity means that of blood erythrocyte.

Catalase Gene

The catalase protein is encoded by a single gene and there is a high of different species - humans, mice, rats, bacteria. The gene is located in position 13 on the short arm of chromosome 11 (Figure 1) near the genes of Wilms tumour, aniridia, and T-cell leukaemia.

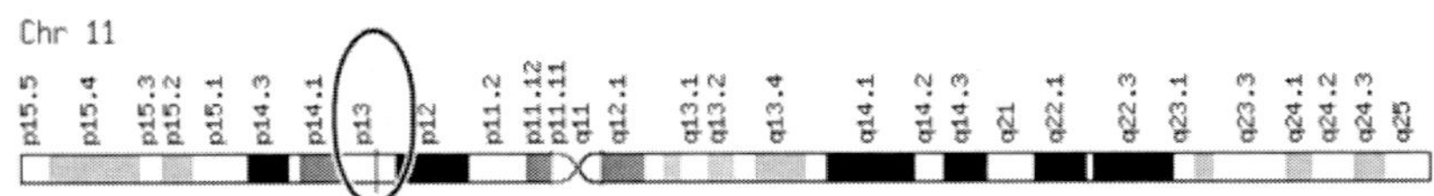

Figure 1. Localization of the catalase gene on chromosome 11.

The gene has 13 exons and 12 introns. The 5' end contains a non-coding region of 68 bp length and a flanking region of 1-320 bp, in which eight transcription sites have been identified. At the end of the gene the 3' flanking region is 747 bp long (Figure 2).

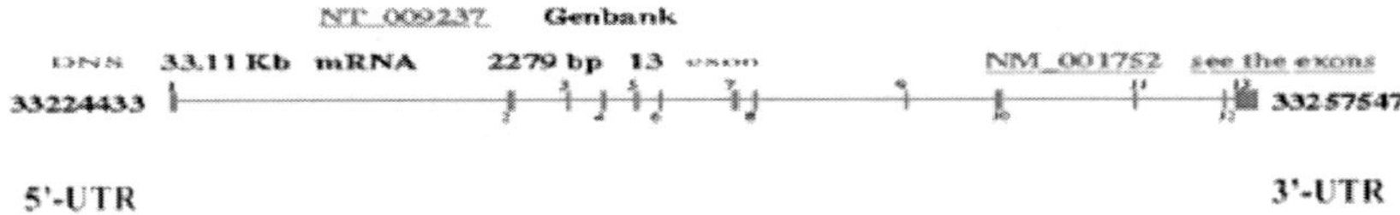

Figure 2. Structure of the catalase gene.

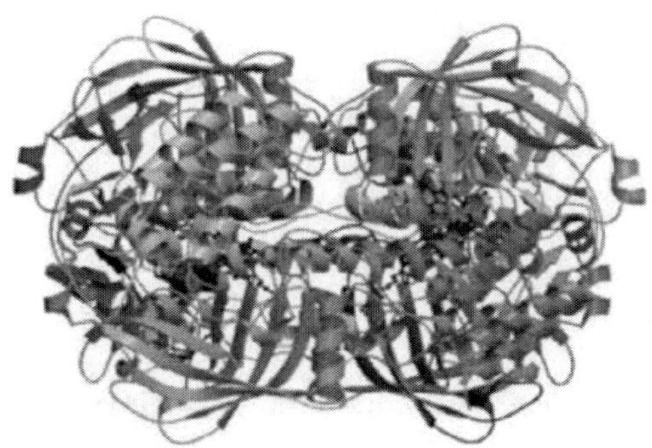

Figure 3. X-ray diffraction structural image of the catalase enzyme.

The synthesis of the catalase protein is on the ribosomes and the subunits with the helper signal protein are transported into the peroxisomes. Hem integration occurs here, followed by dimerization and the tetramer formation (Figure 3).

Inborn Deficiency of Catalase Enzyme, Acatalasemia

Acatalasemia is a homozygous form of congenital catalase deficiency while hypocatalasemia is a heterozygous form of inherited lack of the enzyme. These names reflect the genetic background of the condition. It might be confusing that for acatalesemics the blood catalase activity was below 10%. For explanation, these methods measured the blood catalase activity with the decrease of hydrogen peroxide concentration due to the action of catalase. Contrary, some components in blood could consume hydrogen peroxide too. The heterozygous state yields the enzyme activity of around 50% of those of normal subjects.

To date, 125 individuals with acatalasemia have been reported and they are in 62 families for 12 countries around the world. The majority of patients were detected in Japan (91), while fewer in Switzerland (11) and Hungary (2). In these 3 countries, detailed examinations of the patients concerned and their families have been carried out. Contrary, in other countries there has been no further characterization or investigation of the acatalasemics. Accordingly, there are three types of acatalasemia Japanese, Swiss and Hungarian types that are characterized by detailed biochemical, clinical and, genetic studies. The frequency of acatalsemia is about 0.08/1000 in Japan and 0.05 in Hungary and 0.04/1000 in Switzerland.

Japanese Type of Acatalasemia

The first acatalasemia patient was described in Japan in 1946. This case is one of the first descriptions of congenital enzyme deficiency. Oral surgeon Professor Takahara Seino treated an 11-year-old girl for oral gangrene, periodontitis, and nasal tumours. During the operation, he used hydrogen peroxide for disinfection. Amazing, he did not experience the usual fizziness (oxygen bubble formation) and the blood turned brownish-black. Later, he found the symptoms in 2 more families when he measured their blood catalytic activities. Takahara reported decreased catalase activity in 9 patients of three families, with a complete lack of the enzyme in 1952 [19]. Furthermore, there were 45 patients with a decrease in the activity of the enzyme being heterozygotes of this syndrome. The catalase deficiency was traceable for several generations and in the three families the parents were first or second-degree cousins. Oral gangrene was associated with acatalasemia in most patients, so it was known as Takahara disease in the early papers on acatalasemia. This symptom could be caused by the oxidative tissue-damaging effect of hydrogen peroxide produced by haemolytic staphylococcus and pneumococcus in the mouth.

In the mid-1950s it was detected only sporadically and later it was not found in the newly detected Japanese acatalasemic patients. Poor eating conditions and a lack of basic oral hygiene may be the causes of this symptom.

A genetic study of Japanese cases of acatalasemia identified 2 types of the disease. For the Japanese type A, the G-A splicing mutation at position 5 of the intron 4 causes a decrease in enzyme synthesis. In the case of Japanese type B, deletion of T in nucleotide position 358 of exon 4 causes a frameshift mutation. The electrophoresis of the enzyme molecule shows no change when compared to the controls with normal blood catalase activities [20].

Swiss Type of Acatalasemia

Takahara's findings prompted Swiss biochemistry professor Aebi in the early 1960s to join the research on catalase deficiency. It was not clear for him that Japanese patients revealed any specific clinical sign of the acatalasemia. Therefore, Aebi designed a simple screening test for detection of blood catalase activity during the routine health examination of the army conscripts. In case of reduced catalase activity, they examined the family members too. The study included of more than 18,000 individuals and revealed 11 cases of acatalasemia in 3 Swiss families. These patients were without of oral gangrene (Takahara disease) and any health problems.

There have been only a very limited study of Swiss-type catalase mutations. This is most likely due to the early death (1983) of the program coordinator H. Aebi [21].

Crawford suggested that a regulator mutation might be responsible for a truncated catalase protein [22]. Early studies from Aebi showed that this truncated protein is less stable. This may explain why the catalase deficiency in the erythrocytes of long lifespan is more severe than in other cells of short lifespan [21].

Electrophoretic mobility and isoelectric point of a heterozygote with Swiss type of acatalasemia differed from those of Hungarian acatalasemics and Hungarian normocatalasemic subjects [23]. This phenomenon was not detected in either Japanese or Hungarian patients with acatalasemia.

Hungarian Types of Acatalasemia

In 1989, Goth L. et al. started searching for congenital catalase deficiency for the first time in Hungary. The program lasted for 4 years and the large-scale survey involved clinical, clinical chemistry, and molecular genetic testing of more than 22,000 individuals. It was continued from 1995-2013. This program yielded 14 families with inherited catalase deficiencies. Of these, 13 hypocatalasemic families with a total of 58 heterozygous persons and 68 family members with normocatalasemia. The one acatalasemic family had two homozygotes, four heterozygous, and seven normocatalasemic family members in two generations.

Type A

For the exon 2 at the position of 138 is a GA insertion increasing the GA repeat number from 4 to 5. This GA insertion caused a frameshift in the amino acid sequence from position 68 to 133 and generated a TGA terminating codon at amino acid position 134. Due to this mutation, a trunked protein is formed with a length of 134 amino acids instead of the active protein consisting of 527 amino acids. This mutation was present in the two acatalasemics and 6 heterozygotes of this family. Furthermore, in three hypocatalasemic families, 23 heterozygotes of this mutation were detected [24].

Type B

This type of Hungarian acatalasemia also produces an inactive enzyme. The nucleotide sequence analyses showed a G insertion at position 79 in exon 2. This mutation caused a frameshift in amino acid sequence from 49 to 57 and a TGA stop codon was generated at position 58. This truncated protein with its 58 amino acids compared to the 517 of the regular catalase protein is not able to maintain the enzymatic function. This mutation was detected in three heterozygotes of a Hungarian family [25].

Type C

A T to G substitution at position 5 of intron 7 was detected in 7 members of two hypocatalsemic families. The effect of this splice site mutation was confirmed by Western blot analyses demonstrating a decreased catalase protein level in these patients [26].

Type D

The blood catalase assay of the family revealed decreased activity for 4 family members and normal ones for its 6 members in two generations. The nucleotide analyses showed three intron sequence variations, namely G to A at nucleotide 60 position in intron1, T to A at position 11 in intron2, and G to T at position 32 in intron 12. These variants are unlikely to be responsible for the decreased blood catalase activities as they could be found in the normocatalasemic family members.

However, the novel G to A mutation at position 5 of exon 9 changes the essential amino acid Arg354 to Cys 354. Arg 354 with His 217 probably reveals hydrogen bonding and they could play a role in the catalytic activity of the enzyme and stabilize the electrostatic field [27].

Type E

C to T substitution at position 37 of exon 9 changed Arg365Cys. This mutation causes structural changes to the tertiary conformation of the enzyme. Displacement of the guanidine group causing loss of H-bonding from Arg365, leaving unneutralized charges on the propionic acid carboxylate group that may disrupt the control of the redox potential.

This mutation was detected in two hypocatalasemic family members in an eastern Hungarian family. These patients were suffering from vitiligo. This mutation could contribute to the symptoms of vitiligo but may not be the disease-causing mutation as the other 73 vitiligo patients were without this mutation [28].

Type F1

A T to A nucleotide change at the position of 161 in exon 2 was found. This substitution caused.

Asp54 into Glu yielding changes in the bounds of heme. This mutation was found in one female patient with type 2 diabetes mellitus [29].

Type F2

A G to C substitution at position of 201 in exon 2 turns Glu68 into Asp68 causes decreased blood catalase activity. It was detected in male patient suffering in type 2 diabetes mellitus [29].

Type G

This type of Hungarian acatalasemia was caused by a C insertion in exon 2 (c.106.170insC). This insertion causes a freamshift change from amino acid 35Gly to a TGA stop codon at position 70 (p.G3Afs*5). This truncated catalase protein with its 70, instead of 527, amino acids lacks enzymatic functionality.

This mutation was detected in 2 patients with microcyte anemia and in one patient with type 2 diabetes mellitus [30].

Type H1

As a result of the missense mutation c.379C>T, the CGG triplet becomes TGG (p.R127Y), which results in an amino acid exchange (p.Arg127Tyr). This mutation may decrease the activity of catalase by changing the shape of the main channel and decreasing access of the substrate to the active center. This mutation was identified in a female patient with microcytic anaemia and one with gestational diabetes mellitus [30].

Type H2

The result of the missense c.390T>C mutation, nucleotide change from CGT to CGC is p. Arg129Leu (p.R129L). This nucleotide change of exon 4 may also effect the shape of the main channel and decrease heme affinity for the catalytic His92 residue. It was identified in a male patient with type 2 diabetes [30].

Type H3

The c.431A>T missense mutation changes the triplet of GAT to GTT in exon 4 and turns Asp145 into Val145 (p. Asp145Val, pN145V). Ser198 and Arg144 act as a gate to the narrow part of the main channel. The neighbouring amino acid of Arg114 is reside Asp143.The substitution of Asp143 byVal143 may change the gate and cal lead to the decreased catalase activity. It was identified in a sample of a male patient with type 2 diabetes mellitus [30].

Acatalasemia and Diseases

Homozygous Form

The 113 cases of acatalasemia with 56 males and 57 females are located in Japan, Switzerland Korea, Israel, Peru, Austria, and Hungary.

Acatalsemia was considered "a rare genetic abnormality, asymptotic, no disease" by Takaharain 1977 [20], "no particular clinical importance, acatalasemics are in good health, it could not be considered a disease" by Aebi in 1968 [21] and acatalasemics have no associated health problems" by Eaton and Mu in 1995 [31].

In the absence of particular clinical signs, acatalasemia was generally considered a benign enzyme deficiency until 2000 when a paper appeared reporting an association between humans with inherited catalase deficiency and diabetes mellitus [32]. This was the first finding suggesting that this inherited enzyme deficiency may not be a benign disorder it was generally assumed to be.

In acatalasemic patients of Japan Takahara found that oral lesions such as gangrene, ulceration, and gingival necrosis were often associated with acatalasemia in the late 1940s and early1950s.This symptom might be due to the poor oral hygiene of that time. Later, when oral hygiene improved the prevalence of Takahara diseases decreased and ceased.

Oral lesions were also reported in the cases of Peru and Austria in 1978 and 1999. Of the remaining cases of acatalasemia, one patient had a dermatological problem, one had a disc problem, two had diabetes mellitus, and one had laryngeal carcinoma. The eleven Swiss acatalasemics had no health problems reported [33].

Heterozygous Form

Patients with the heterozygous form of inherited catalase deficiency yielded their blood catalase activity between 44.9% and 74.0% compared to those of normocatalasemic family members. Their diseases were Parkinson's disease for one patient, laryngeal cancer for another one, and 313 patients were on normal life.

The eleven types of catalase gene mutations of the Hungarian patients were associated with decreased blood catalase activities (53.1±13,4 MU/L, n:

53, 48.4%) when compared to those of normocatalasemic family members (109,7±13,3, n:45). The mutations were detected in exon 2 (types A, B, F1, F, G), in exon 4 (H1,H2,H3), in exon 9 (D,E), in intron 1 (D) and in intron 7 (C).

Diseases of the Hungarian heterozygous patients were diabetes mellitus for 11 patients, arteriosclerosis for 4 patients, microcytic anemia for 3 patients, vitiligo for 2 patients, schizophrenia for one patient and 32 patients had no associated health problems living a normal life.

The oxidative stress due to increased hydrogen peroxide production in catalase deficiency could contribute to the manifestation of diabetes. Furthermore, for some ROS diseases such as microcytic anemia, arteriosclerosis, vitiligo, and schizophrenia it may be one of the factors in their causations.

Inherited catalase deficiency is associated with clinical features, pathologic laboratory test results, age and oxidative stress-related disorders. Rather than considering it a benign condition, it should be considered as a condition for aging and oxidative stress [33].

Diabetes Mellitus

Diabetes mellitus is a metabolic disorder characterized by hyperglycemia and either insufficiency of insulin secretion or a lack of insulin receptor function. Multiple genetic and environmental factors may be involved. Micro and macro-vascular complications of diabetes are the most common causes of renal failure, blindness, and amputations, which lead to significant mortality, morbidity, and poor quality of life. According to the International Diabetes Federation (IDF), 537 million adults (20-79 years old) are affected, but based on the risk group, the number of patients could reach 783 million by 2045.

Type 2 diabetes mellitus is considered a complex disease and its development is influenced by both environmental factors and genetic factors.

Hur and colleagues conducted literature research in 2010 to look for and detect genes that are associated with diabetes and reactive oxygen agents. Using the SciMiner program, they found that in the relation between ROS and diabetes mellitus, insulin is first, the second is superoxide dismutase, and the third is catalase [34].

Blood Catalase and Diabetes Mellitus

Blood catalase activity of diabetic patients was performed in four studies.

For study one a total of 137 randomly selected diabetics (type 1: 45, type 2: 92) yielded a significant (P:0.001) decrease in their blood catalase (94.4±19.2 MU/L) when compared to that of controls (113.3±16.5 MU/L) [35].

Study 2 was related to pregnant diabetics. The mean activity of blood catalase was decreased (P0.001) in pregnant patients without diabetes (89±18 MU/L,n:169) when that of control females was 109±13 MU/L (n:235). For diabetic pregnant the blood catalase showed a significant (P0.001) decrease (74±14MU/L, n:60) when compared to either normal females or to pregnant without diabetes [36].

Study 3 examined another cohort of diabetics with type 1(106), type 2 (100), gestational (33), and controls (60). Compared to the controls (104.7±18.5 MU/L) the blood catalase activity decreased (P0.001) in type 2 diabetes (71.2±14.6 MU/L) and gestational diabetes (68.5±12.2 MU/L) and it was unchanged (P>0.05) in type 1 diabetes mellitus (102.5±26.9 MU/L) [37].

Study 4 included a further cohort of diabetic patients (type 1:115, type 2: 225) and controls: (295). Its results were similar to those of cohort 3. Blood catalase activity of type 1 diabetics was not changed (101±24 MU/l, P>0.05) while type 2 diabetics had significantly (P0.001) lower blood catalase (91±22 MU/L) compared either to type 1 diabetics (101±24 MU/L) or to controls (104±15 MU/L) [38].

These studies clearly showed that blood catalase activity decreases in type 2 and in gestational diabetes while it does not change in type 1 diabetes. This may be due to either catalase gene mutations or regulatory mechanisms.

Examination of possible catalase gene mutations which could be responsible for the significant decreased blood catalase activity was performed in 380 diabetics (type 1:115, type 2:205, gestational:60). Highly decreased blood catalase was detected in 23 patients (6.0%) with type 1:1, type 2:14 and gestational diabetes: 3. Their blood catalase activities were less than 52 MU/L which means the 50% of the mean (52 MU/L9 of the controls (104±15 MU/L, n: 295).

Their mutation screening and nucleotide sequence analyses revealed four acatalasemia mutations one for the gestational diabetic (H1), and three for type 2 diabetics (H2, H3 and G). These results can suggest the association between acatalasemia mutations and type diabetes mellitus (OR:144.8, 0.95CI: 21.5-974.1) and gestational diabetes (OR: 144.8, 0.95CI:10.2-2059-6).

Furthermore, these acatalsemia mutations could explain only the decreased (less than 50%) blood catalase activities in 21.4% (3/14) of type 2 diabetics and 33.3% (1/3) of gestational diabetics [29].

In type 2 diabetes the ratio of females to males has a slightly significant change (P0.0007) in inherited catalase deficiency (female: 11, male: 2, ration 5.5) when compared to that of type 2 diabetics without acatalasemia mutations and normocatalasemia (female: 80, male: 134, ration: 0.59).

When normocatalasemic subjects (n:18, age:44.7±17.9 years, blood catalase: 101±9 MU/L) were compared to their age– and sex-matched hypocatalasemic family members (n:18, age: 45.1±17.9 years, blood catalase 61±10 MU/L), increased glucose concentrations for hypocatalasemics (hypocatalasemia: 5.42±0.8 mmol/L vs normocatalasemia: 4.83±0.46 mmol/l) were received. The higher average glucose (5.42 mmol/l) of hypocatalasemic subjects may indicate that they have a larger (3.05x) risk for type 2 diabetes than their normocatalasemic (glucose: 4.83 mmol/l, risk: 1.81x) family members.

The onset of type 2 diabetes for catalase deficient patients (43.1±10.9 years) appeared more than 10 years earlier (P0.001) than for normo-catalasemic diabetic subjects (56.3±11.2 years).

For inherited acatalasemia the 2 acatalasemics (2/2) and 11 hypocatalasemics (11/63) had diabetes mellitus. One male patient had type 1 while 10 were with type 2 forms. The prevalence of type 2 diabetes in Hungarian patients with inherited catalase deficiency is 15.9%. In contrast, the prevalence of type 2 diabetes was 0% among 65 normocatalasemic family members and about 7% among Hungarians in general.

In conclusion, these data revealed an association between diabetes mellitus especially its type 2 with inherited catalase deficiency. One of unique features of beta cells of pancreas is their relatively low expression of many oxidant enzymes including catalase. Furthermore, while beta-cells are also rich in mitochondria, the major source of endogenous generation of superoxide and hydrogen peroxide. Therefore, it may be postulated that an increased level of hydrogen peroxide stress, due to decreased catalase activity, extending over a period of years might result in the slow accumulation of oxidant damage to the beta-cells and full-blown diabetes [29, 30, 32, 39, 40].

Catalase Past and Future

Enzyme catalase was first detected in 1819, first used in clinical laboratory practice and tumor research in1910 [1], first reported as an inherited enzyme deficiency in 1948 [20], and on its enzyme-substrate complex in 1948 [41].

For hydrogen peroxide metabolism the role of catalase was summarized in 1979 [13]. In the first years of this century its new role in signaling has been examined [42].

The next titles of papers show the unbroken interest in catalase research. "Catalase: an old enzyme with a new role? "by Percy in 1984 [43], "Catalases and peroxidases and glutathione and hydrogen peroxide. Mysteries of the bestiary" by Eaton in 1991 [44], "Mammalian catalase a venerable enzyme with new mysteries" by Kirkman in 2006 [45], "Catalase, a remarkable enzyme: targeting the oldest antioxidant enzyme to find a new cancer treatment approach" by Glorieux in 2017 [1], "The venerable catalase will continue to bewilder us with its mysteries well into the twenty-first century" by Sepsi 2018 [46] and "A new paradigm in catalase research" by Fujiki in 2021 [47].

References

[1] Glorieuxa Christopher, Calderon Buc Pedro. (2017) "Catalase, a remarkable enzyme: targeting the oldest antioxidant enzyme to find a new cancer treatment approach. *Biological Chemistry* 398(10):1095-1108.

[2] Loew Oscar. (1901) "Catalase a new enzyme of general occurrence with reference to the tobacco plant." *US Department of Agriculture Report* 68: 47.

[3] Bergmeyer Hans Ulrich. (1955). "Zur Messung von Katalase-Aktivitater." *Biochemische Zeitschrift* 327:255.

[4] Aebi Hugo. (1984) "Catalase in vitro." *Methods in Enzymology.* 105:121-126.

[5] Beers Roland F, Sizer Irwin W. (1952) "Spectrophotometric method for measuring breakdown of hydrogen peroxide by catalase" *Journal of Biological Chemistry* 195:133-140.

[6] Sinha Asru K. (1972) "Colorimetric assay of catalase" *Analytical Biochemistry* 47: 389-394.

[7] Goth Laszlo. (1991) "A simple method for determination of serum catalase activity and revision of reference range" *Clinica Chimica Acta* 196:143-152.

[8] Johansson Lars H, Borg Hakan LA. (1988) "A spectrophotometric method for determination of catalase activity in small samples" *Analytical Biochemistry* 174:331-336.

[9] Vitai Marta, Goth Laszlo. (1997) "Reference ranges of normal blood catalase activity and levels in familial hypocatalasemia in Hungary" *Clinica Chimica Acta* 261:35-42.

[10] Veal Elisabeth. Day Alison. (2011) "Hydrogen peroxide as a signaling molecula" *Antioxidants and Redox Signaling.*15:147-151.

[11] Rajendran, P., Nandakumar, N., Rengarajan, T., Palaniswami, R., Gnanadhas, E. N., Lakshminarasaiah, U., Gopas, J., & Nishigaki, I. (2014) "Antioxidants and human diseases" *Clinica Chimica Acta* 436:332–347.

[12] Halliwell, B., Clement, M. V., & Long, L. H. (2000) "Hydrogen peroxide in the human body" *FEBS Letters* 486:10-13.

[13] Chance, B., Sies, H., & Boveris, A. (1979) "Hydroperoxide metabolism in mammalian organs" *Physiological Review.* 59:527-605.

[14] Wakimoto, M., N. Masuoka, T. Nakano, T. Ubuka. (1998) "Determination of glutathion peroxidase activity and its contribution to hydrogen peroxide removal in erythrocytes." *Acta Medica Okayama* 52:233-37.

[15] Rhee Sou, Chaez Ho Zoon, Kim Hojin. (2005) "Peroxiredoxins: a historical overview and speculative preview of novel mechanism and emerging concepts in cell signaling." *Free Radicals in Biology and Medicine.*38(12):1543-1552.

[16] Góth Laszlo. (1982) "Determination of catalase enzyme activity in human tissues by programmable polarograph." *Hungarian Scientific Instruments* 53:43-46.

[17] Olafsson Tor, Olafsson Inge. (1977) "Purification of human granulocyta catalase in chronic myeloid leukemia" *Biochemica et Biophysica Acta Enzymology.* 482:301-308.

[18] Seghieri, G., Di Simplicio, P., Anichini, R., Alviggi, L., De Bellis, A., Bennardini, F., & Franconi, F. (2001) "Platelet antioxidant enzymes in insulin-dependent diabetes mellitus" *Clinica Chimica Acta* 309:19-23.

[19] Takahara Seino. (1952) "Progressive oral gangrene due to lack of catalase in the blood (acatalasemia)." *Lancet* 2:1101-1104.

[20] Takahara Seino, Ogata Masana. (1977) Metabolism Japanase acatalasemia wit special reference to superoxide and glutathione peroxidise. In *Biomedical and Medical Aspects of Oxygen,* edited by Hayaishi O, Asada K., Baltimore, University Park Press 275-292.

[21] Aebi Hugo, Wyss Sonja. (1977) *Acatalasemia.* In *The metabolic and molecular basis of inherited diseases.* edited by Stanbury J. B., Wingaarden J. B., Frederickson 4th ed. New York, Mc Graw Hill 1972-1807.

[22] Crawford, D. R., Mirault, M.-E., Moret, R., Zbinden, I., & Cerutti, P. A. (1988) "Molecular defect in human acatalasia fibroblast." *Bichemical and Biophysical Research* Communications 153:59-66.

[23] Goth Laszlo. (1992) "Characterization of acatalasemia detected in two Hungarian sisters " *Enzyme* 46:252-258.

[24] Góth, L., Shemirani, A., & Kalmár, T. (2000) "A novel catalase mutation (a GA insertion) causes the Hungarian type of acatalasemia." *Blood Cells Molecules and Dieasess.* 26:151-154.

[25] Goth Laszlo. (2001) "A novel catalase mutation (a G insertion in exon 2) causes the type B of the Hungarian acatalasemia." *Clinica Chimica Acta.* 311:161-163.

[26] Goth Laszlo. (2001) "A new type of inherited catalase deficiencies: Its characterization and comparison to the Japanese Mad Swiss type of acatalasemia." *Blood Cells Molecules and Diseases.* 27(2):512-517.

[27] Góth, L., Vitai, M., Rass, P., Sükei, E., & Páy, A. (2005) "Detection of a novel familial catalase mutation (Hungarian type D) and the possible risk inherited catalase deficiency for diabetes mellitus." *Electrophoresis* 26: 1646-1649.

[28] Kósa, Z., Fejes, Z., Nagy, T., Csordás, M., Simics, E., Remenyik, É., & Góth, L. (2012) "Catalase -262C to T polymorphisms in Hungarian vitiligo patients and in controls: further acatalasemia mutations in Hungary." *Molecular Biology Reports* 39: 4787-4795.

[29] Góth, L., T. Nagy, G. Paragh, M. Káplár. (2016) "Blood catalase activities, catalase gene polymorphisms and acatalasemia mutation in Hungarian patients with diabetes mellitus." *Global Journal of Obesity, Diabetes and Metabolic Syndrome* 3(1):1-5.

[30] Nagy, T., Paszti, E., Kaplar, M., Bhattoa, H. P., & Goth, L. (2015) "Further acatalasemia mutations inhuman patients from Hungary with diabetes and microcytic anemia." *Mutation Research/Fundamental and Molecular Mechanisms of Mutagenesis.* 772:10-14.

[31] Eaton John W, Mu M. (1995). Acatalasemia. In *The Metabolic and Molecular Basis of Inherited Diseases.* edited by Sciver CR, Beaudet A 2nd ed. New York, Mc Graw Hill 2371-2384.

[32] Goth L, Eaton John W. (2000) "Hereditary catalase deficiency and increased risk of diabetes." *Lancet* 356:1820-1821.

[33] Goth Laszlo, Nagy Terez. (2013) "Inherited catalase deficiency: Is it benign or a factor in various age related diseases. *Mutation Research/Reviews in Mutation Research* 753:147-154.

[34] Hur, J., Sullivan, K. A., Schuyler, A. D., Hong, Y., Pande, M., States, D. J., Jagadish, H. V., & Feldman, E. L. (2010) "Literature-based discovery of diabetes- and ROS-related targets" *BMC Medical Genomics* 3:49-60.

[35] Góth, L., Lenkey, A., & Bigler, W. N. (2001) "Blood catalase deficiency and diabetes in Hungary." *Diabetes Care.* 24:1839-1840.

[36] Góth, L., Tóth, Z., Tarnai, I., Bérces, M., Török, P., & Bigler, W. N. (2005) "Blood catalase activity in gestational diabetes is decreased but not associated with pregnancy complications." *Clinical Chemistry.* 51:2401-2404.

[37] Tarnai, I., Csordás, M., Sükei, E., Shemirani, A. H., Káplár, M., & Góth, L. (2007) "Effect of C111T polymorphism in exon 9 of the catalase gene on blood catalase activity in defferent types of diabetes mellitus." *Free Radical Research.* 41:806-811.

[38] Góth, L., Nagy, T., Kósa, Z., Fejes, Z., Bhattoa, H. P., Paragh, G., & Káplár, M. (2012) "Effects of rs769217 and rs1001179 polymorphisms of catalase gene on blood catalase, carbohydrate and lipid biomarkers in diabetes mellitus." *Free Radical Research* 46(10):1249-1257.

[39] Goth Laszlo, Nagy Terez. (2012) "Acatalasemia and diabetes mellitus." *Archives of Biochemistry and Biophysics* 55:195-200.

[40] Góth, L., Rass, P., & Páy, A. (2004) "Catalase enzyme mutations and their association with diseases." *Molecular Diagnosis* 8(3):141-149.

[41] Britton Chance. (1948) "The enzyme-substrate compounds of catalase and peroxides." *Nature* 161:914-917.

[42] Rhee Sue Goo. (2006) "H_2O_2, a necessary evil for cell signaling." *Science* 312:1882– 1883.

[43] Percy Maire E. (1984) "Catalase: an old enzyme with a new role?" *Canadian Journal of Biochemistry and Cell Biology* 62(10):1006-10014.

[44] Eaton John W. (1991) "Catalases and peroxidases and glutathione and hydrogen peroxide. Mysteries of the bestiary" *Journal of Laboratory and Clinical Medicine* 117:3-4.

[45] Kirkman, H. N., & Gaetani, G. F. (2006)." Mammalian catalase a venerable enzyme with new mysteries" *Ternds in Biochemical Sciences* 32(1):44-50.

[46] Tehrani Hessam Sepsi, Moosavi-Movahedi Ali, Akbar. (2018) "Catalase and its mysteries." *Progress in Biophysics and Molecular Biology.* 140:5-12.

[47] Fuyiki Yukio Yukio, Bassik Michael. (2021) "A new paradigm in catalase research" *Trends in Cell Biology* 31(3):148-151.

Chapter 5

Catalase and Its Inhibition: Inhibitory Effects of Drugs on Catalase Activity

Safija Herenda[1,*], PhD and Edhem Hasković[2], PhD
[1]Department of Chemistry, Faculty of Science, University of Sarajevo, Sarajevo, Bosnia and Herzegovina
[2]Department of Biology, Faculty of Science, University of Sarajevo, Sarajevo, Bosnia and Herzegovina

Abstract

Catalase belongs to the group of oxidoreductases, and as an antioxidant enzyme, it plays a major role in eliminating hydrogen peroxide, a by-product of many normal cellular reactions. Reduced catalase activity increases oxidative stress in the body. The link between oxidative stress and many diseases suggests that each endogenous catalase inhibitor has the potential to contribute to the occurrence of certain pathological conditions. A non-functional antioxidant system with reduced catalase activity in the cell can lead to pathological conditions such as certain types of cancer, Alzheimer's disease, diabetes, including inflammatory processes. Certain drugs given for therapeutic purposes can reduce the activity of catalase in the cell, so in this chapter, we will talk about the kinetics and mechanism of catalase inhibition. Drugs exert various effects on catalase, acting as competitive, non-competitive or a competitive enzyme inhibitors. In this process, the drug molecule undergoes a chemical transformation and becomes a suitable product that blocks the normal flow of the metabolic pathway. Inhibitors specifically bind to enzymes and modify the enzymatic reaction rate. The same substance in relation to the substrate concentration can be both an

* Corresponding Author's Email: islamovic.safija@gmail.com.

In: Catalase and Its Applications
Editor: Kaley Rutherford
ISBN: 979-8-88697-421-8
© 2022 Nova Science Publishers, Inc.

activator and an inhibitor of the test enzyme. Due to structural changes resulting from substrate-inhibitor binding, partial inhibitors may also be allosteric inhibitors of enzyme activity. By determining the kinetic constants of the Michaelis-Menten constant, the maximum rate of the enzymatic reaction, the catalytic constant and the inhibition constant, information can be obtained on the enzyme catalysis efficiency and the mechanism of catalase inhibition itself.

Keywords: catalase, inhibition, drugs, enzyme, inflammation

Introduction

Enzymes are biocatalysts that accelerate chemical reactions inside living cells, during which they remain unchanged. Each enzyme is specific for a particular substrate or substrates, during which it produces a specific product. According to their chemical structure, they belong to proteins, and their primary characteristic is catalytic activity. The site responsible for the catalytic activity is called the active site and is part of the enzyme's tertiary structure. It is often a hydrophilic cleft or cavity with a series of amino acid side chains that carry out the enzymatic reaction by binding the substrate. The active site makes up only 10-20% of the total mass of the enzyme.

Enzymes are composed of polypeptide chains of amino acids that bend, forming hydrogen bonds and creating a secondary structure. Further twisting of the chains forms the tertiary structure of the enzyme. This final form of the enzyme represents its conformational form, in which the essential amino acids and cofactors are in a specific spatial orientation relative to each other, which enables the enzyme to perform catalysis. The quaternary level of structural organization has enzymes built from at least two polypeptide chains or subunits. Enzymes control and regulate all physiological processes and are considered necessary for any metabolic processes in the cell (Fontes, Ribeiro and Sillero, 2000). They exhibit their catalytic activity in small quantities, at relatively low pressure, and emerge from the reaction unchanged. The International Union of Biochemistry (IUB) introduced the nomenclature of enzymes in 1964, where enzymes are mainly divided according to the specificity of enzymatic reactions into six large groups: oxidoreductases, transferases, hydrolases, lyases, isomerases and ligases. EC: Enzyme Commission Number is each enzyme's designation in the classification overview, based on the chemical reaction they catalyze. Of particular importance in the study of enzyme action mechanisms are the class of enzymes that regulate the level of oxidative stress at the level of the cell or organism, namely oxidoreductases. The impaired activity of these enzymes creates a predisposition to the emergence of various pathological conditions, especially cancer. Current therapies used for cancer treatment, in addition to certain benefits, also have proven harmful effects, i.e., they destroy diseased and healthy cells. Today, substantive work is being done to find and create drugs targeting only diseased cells while healthy cells remain intact. Potential drugs would modify the action of oxidoreductases inside the cell, manifesting their inhibitory properties (Copeland, 2005).

Literature Review

The general, simplest enzymatic reaction can be written as:

$$E + A \underset{k_{-1}}{\overset{k_1}{\rightleftharpoons}} EA \xrightarrow{k_2} E + P \tag{1}$$

The substrate, marked as A, reacts with the enzyme E, forming a substrate-enzyme complex EA, which is converted into product P by an irreversible reaction. The unchanged enzyme is released, entering a new reaction.

Reaction degrees are determined by constants k. Negative constants indicate that the reaction moves in the opposite direction (Marangoni, 2003). However, enzymatic reactions are much more complex in practice and involve several components such as one or more substrates and products, cofactors, inhibitors and activators. During the interaction between enzyme and substrate, we distinguish:

1. *Electrostatic interactions*. Substrates containing functional groups that ionize in an aqueous solution are often electrostatically bound to the oppositely charged amino acid side chains that form the enzyme's active site. The energy of this binding is in the range of 25 to 50 kJ mol^{-1}.
2. *Hydrogen bonds*. This interaction is common when binding functional groups of polar substrates, where the strength of this bond depends on the chemical nature and geometry of the interacting groups.

3. *Van der Waals interactions.* Given that the active site is complementary to the substrate, this interaction has high binding energy, but individual reactions are weak.
4. *Hydrophobic interactions.* Suppose the substrate has a hydrophobic surface or contains a hydrophobic group. In that case, binding can only occur if the substrate is bound in the hydrophobic part of the active site of the enzyme (Bugg, 2012).

Catalase

Oxidoreductases catalyze the transfer of H atoms, O atoms or electrons from one substrate to another. Oxidoreductases belong to the first class of enzymes, denoted by EC 1. In this class, numerous subclasses are present, the number of which tells us what type of donor is involved during the reaction. Some oxidoreductases are catalase, dehydrogenases, reductases, and peroxidases (Palmer and Bonner, 2007). Catalase is an enzyme widely distributed in the living world from prokaryotes to eukaryotes, aerobic microorganisms, plants, and animals, that is, in almost all aerobic organisms. In humans, catalase activity is expressed in peroxisomes of hepatocytes and erythrocytes. Catalase has a role in the antioxidant protection of the body, that is, to protect cells from the toxic effects of hydrogen peroxide (Milton, 2008). Catalase is a homotetramer consisting of four polypeptide chains. Individual polypeptide chains are made up of 572 amino acids. It has two cofactors: heme group and NADP(+) and four porphyrin hemes. The catalase reaction with hydrogen peroxide is facilitated by prosthetic heme groups and promotes the growth of cells, including T- and B-lymphocytes (Milton, 2008). Catalases show the highest enzymatic activity in the liver and erythrocytes. In cells, they are mainly found in peroxisomes and mitochondria as soluble and membrane-bound forms (Đorđević, 2004). Several organic substances, e.g., ethanol, can act as hydrogen donors. Heme-dependent catalases form the Fe4+ = O porphyrin π-cation radical as an intermediate (Purich and Allison, 2002). Catalase catalyzes the conversion of two molecules of hydrogen peroxide into two molecules of water and one molecule of oxygen, according to the reaction:

$$2H_2O_2 \rightarrow O_2 + 2H_2O \tag{2}$$

The mechanism of action of catalase has not been fully resolved; however, a two-step mechanism has been proposed:

$$H_2O_2 + Fe(III) - E \rightarrow H_2O = Fe(IV) - E \quad (3)$$

$$H_2O_2 + O = Fe(IV) - E \rightarrow H_2O + Fe(III) - E \quad (4)$$

Fe-E represents the heme iron center attached to the rest of the enzyme. After peroxide enters the heme cavity, due to steric hindrances, it reacts with amino acids His^{74} (histidine) at position 74 and Asn^{147} (asparagine) at position 147. In this position, the first phase of catalysis begins. During the transfer of a proton from one peroxide oxygen atom to another through His^{74}, the bond between the two oxygen atoms is polarized and finally heterolytically cleaved while the peroxide is coordinated to the iron center. This coordination releases a water molecule and forms a Fe(IV) = O complex and a heme radical. The free radical very quickly transfers an electron making it a radical, while the heme ring remains unchanged. In the second phase, similarly by transferring two electrons, the Fe(IV) = O complex reacts with another molecule of hydrogen peroxide, turning into Fe(III)-E and producing one molecule of water and one mole of molecular oxygen. It is assumed that the efficiency of catalase lies in the presence of His^{74} and Asn^{147} and their interaction with intermediates.

Mechanism of Enzyme Catalytic Activity

In chemical reactions, enzymes increase the rate of the reaction in such a way that the reaction approaches an equilibrium state. The enzyme's catalytic activity results from the substrate's high specificity for the substrate, the optimal organization of the catalytic groups that stabilize the transition state at the top of the energy barrier. Catalytic mechanisms of enzymes can be geometric requirements of catalysis, acid-base catalysis, covalent catalysis, metal-ion catalysis, and catalysis by preferential binding of enzymes to the transient state.

The geometric requirements of catalysis imply the establishment of geometrically regular interaction between the enzyme and the substrate. The enzyme binds the substrate so that the reacting chemical bond of the substrate is positioned and oriented close to the catalytic groups of the active site of the enzyme so that the substrate then goes into the transition state. This type of catalysis mechanism is represented by allosteric enzymes that perform catalysis according to the induced fit model. The specific free energy of binding the substrate to the enzyme allows optimal placement of the substrate in the active site of the enzyme, where the free energy of activation is lowered by giving or removing a proton from the substrate. In general acid catalysis, the free energy of activation of the transient state is reduced by the temporary transfer of protons from the proton donor molecule/group (Brnösted acids). In general, base catalysis takes place by temporarily taking over a proton from a Brönsted base (proton acceptor molecule/group). Acid-base catalysis combines two described catalysis mechanisms in different parts of catalysis. The described mechanism of catalysis is present in most metabolic processes.

Covalent catalysis is a mechanism of catalysis in which the substrate is covalently bound to the enzyme, and thus transient covalent intermediates are formed. In the first phase, a covalent bond is formed between the enzyme and the substrate through a nucleophilic reaction. In the second phase, the electrophilic part of the enzyme's active centre withdraws an electron. In the third phase, the enzyme leaves the reaction unchanged.

Metal-ion catalysis is a mechanism of enzyme catalytic activity that requires a metal ion. Enzymes can be metalloenzymes (strong bond between metal ion and enzyme) or enzymes activated by metal ion (weak bond between metal ion and enzyme) according to the strength of interaction between metal ion and enzyme. The metal ion has several roles in catalysis, such as enabling the enzyme-substrate bond formation, establishing the proper orientation of the substrate, participating in oxidation and reduction processes due to the reversible change in the oxidative state of the metal and performing electrostatic stabilization of the substrate.

Catalysis by preferential binding of the enzyme to the transient state is a mechanism of catalysis where the active site of the enzyme has a greater affinity for binding the transient state of the substrate than for the substrate itself or for the product of the reaction. The enzyme first acts mechanically on the substrate and translates the substrate into a transient state, which best fits into the enzyme's active site (Maragoni, 2003).

Types of Enzyme Inhibition

Any change in the K_M value for the same enzyme and substrate often indicates an inhibitor or activator presence. Small K_M values indicate that the enzyme requires a small amount of substrate to become saturated. Large K_M values indicate the need for high substrate concentrations to achieve a maximum reaction rate (Cornish-Bowden, 1995).

The rate of an enzymatically catalyzed reaction is defined as:

$$v_0 = \frac{V_{max}[A]}{K_M+[A]} \quad (5)$$

where A denotes the substrate, Vmax the maximum velocity, and K_M the Michaelis-Menten constant. Determining the kinetic parameters K_M and V_{max} is essential for describing enzyme-catalyzed reactions.

However, since these values cannot be calculated most accurately from the hyperbola, a more appropriate representation of the so-called double reciprocal plot or Lineweaver-Burk plot is presented (Furman, Painter and Anderson, 2000).

$$\frac{1}{v_0} = \frac{1}{V_{max}} + \frac{K_M}{V_{max}} + \frac{1}{[S]} \quad (6)$$

The segment on the ordinate represents the reciprocal value of Vmax, the segment on the abscissa represents the reciprocal value of - K_M, and the slope of the line represents K_M/V_{max}.

Enzymes are, in principle, highly specific to the substrate. Based on this, we distinguish four different types of specificity:

1. *Absolute specificity.* An enzyme catalyzes only one type of chemical reaction, i.e., selectively acts on one specific substrate.
2. *Group specificity.* It is manifested by enzymes that seek only a particular type of bond, those bonds in the construction of which one of the permanent components takes part, e.g., glucosidic bond.
3. *Low specificity.* It is manifested by enzymes that recognize only a certain type of bond: ester, glucosidic, peptide.
4. *Stereochemical specificity.* The enzyme acts individually on a steric or optical isomer (Bisswanger, 2008).

This selectivity is not absolute, so another molecule may bind to the enzyme's active site if it is structurally homologous to the substrate. Knowing the configuration of the enzyme's active site, it is possible to design such a molecule, which in some cases binds to the enzyme even more specifically than the substrate. Most often, this is used to block or deactivate the active site to prevent a certain enzymatic reaction; this is how most pharmaceutical products work.

Competitive Inhibition

This type of inhibition is present if the inhibitor is a compound that is structurally related to the substrate. The substrate and enzyme bind to the active site of the enzyme, forming an enzyme-substrate complex and an enzyme-inhibitor complex. In this way, the binding of the substrate to the enzyme is reduced, and a higher concentration of the substrate is required to reach the maximum speed of the enzyme-catalyzed reaction. This reaction can be shown schematically as:

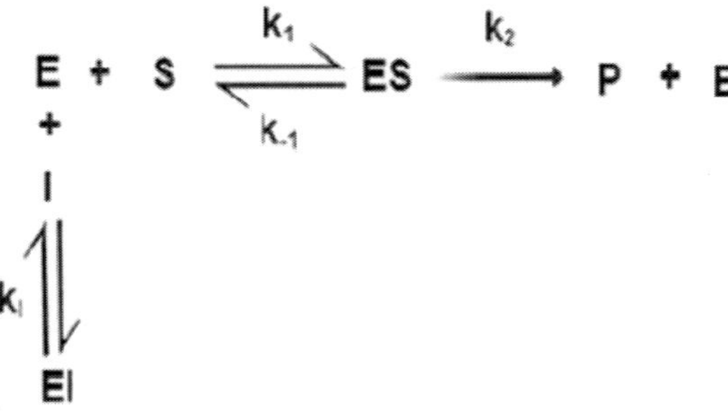

Figure 1. Schematic representation of competitive inhibition.

where E is the enzyme, S is the substrate, and I is the inhibitor (Figure 1). Binding affinities are expressed by dissociation constants K_1 and K_I (Bisswanger, 2008). K_M increases in the presence of a competitive inhibitor, so the section on the ordinate of the Lineweaver-Burk diagram does not change. Still, the slope of the line increases and the section on the abscissa decreases (Figure 2).

$$\frac{1}{v_0} = \frac{K_M}{V_{max}}\left(1 + \frac{[I]}{K_I}\right) + \frac{1}{[S]} + \frac{1}{V_{max}} \quad (7)$$

One example of a competitive enzyme inhibitor is the HIV aspartyl protease inhibitor. The development of HIV aspartyl protease inhibitors has been primarily driven by a combination of mechanism-based and structure-based drug designs. The design of the competitive inhibitor considered known peptide substrates of the enzyme and known transition state mimics of aspartyl proteases (statins, hydroxyethylenes, etc.). Introducing a statin or hydroxyethylene group into the peptide substrate at the scissile amide bond site has produced many viral protease inhibitors. These were reasonable initial steps for inhibitor design but were too peptide in nature to have good pharmacological properties. A significant advance in the design of HIV protease inhibitors began with the report of the enzyme's crystal structure by Navia et al. (Navia, Fitzgerald, McKeever, Leu, et al. 1989).

Noncompetitive Inhibition

In noncompetitive inhibition, the substrate and the inhibitor bind to the enzyme, but in different places where they affect each other due to, e.g., steric or electrostatic interactions (Figure 3).

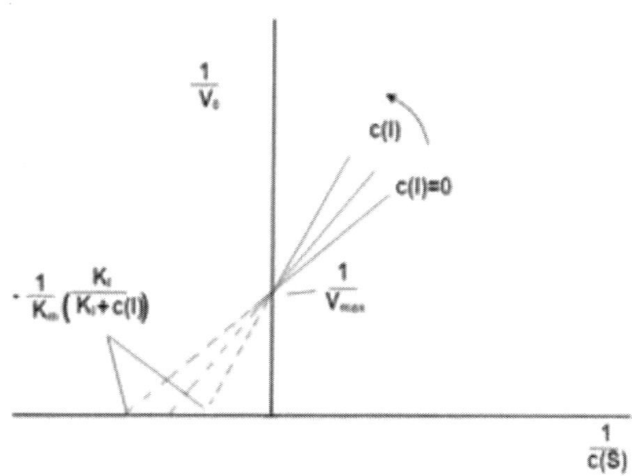

Figure 2. Lineweaver-Burk diagram for competitive inhibition.

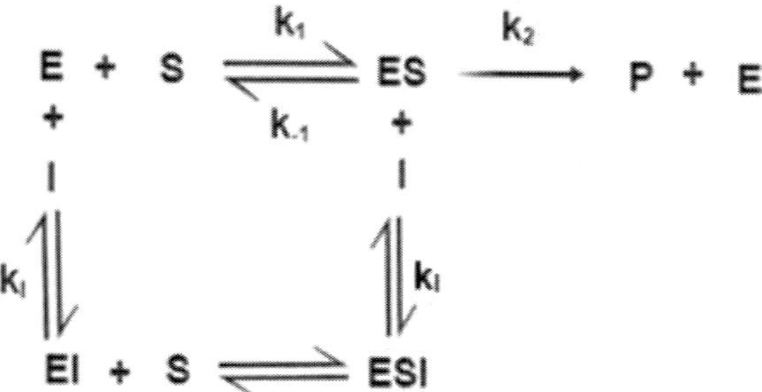

Figure 3. Schematic representation of non-competitive inhibition.

The slope and intercept on the ordinate change on the Lineweaver-Burk plot, i.e., the section on the ordinate is higher in the presence of the inhibitor, which would mean that with the addition of the inhibitor, V_{max} decreases and cannot be increased if the concentration of the substrate changes. K_M, in this case, remains the same because the saturation of the active site does not affect the binding of the inhibitor (Figure 4).

$$\frac{1}{v_0} = \frac{K_M}{V_{max}}\left(1 + \frac{[I]}{K_I}\right)\left(\frac{1}{[S]}\right) + \frac{1}{V_{max}}\left(1 + \frac{[I]}{K_I}\right) \tag{8}$$

Non-nucleoside reverse transcriptase inhibitors (NNRTIs), used in treating AIDS, provide compelling examples of clinically relevant non-competitive inhibitors. The causative agent of AIDS, HIV, belongs to a family of viruses that rely on the genetic system. Viral replication requires reverse transcription of the viral genomic RNA into DNA, which is then incorporated into the genome of the infected host cell. Reverse transcription is catalyzed by a virally encoded nucleic acid polymerase known as reverse transcriptase (RT). This enzyme is critical for viral replication, and inhibition of HIV RT is an effective mechanism for reducing infection in patients (Christopoulos, A. 2002).

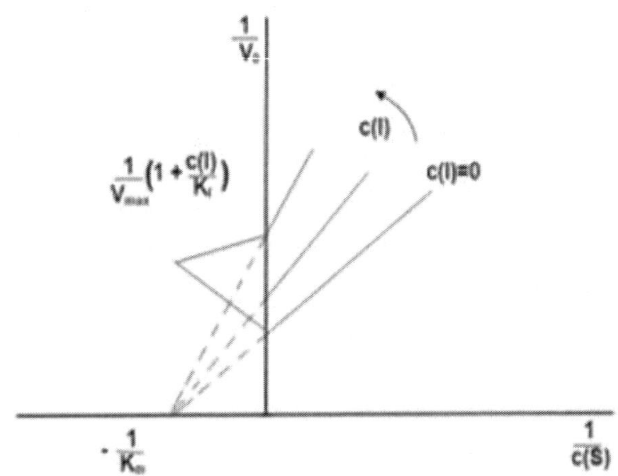

Figure 4. Lineweaver-Burk plot for non-competitive inhibition.

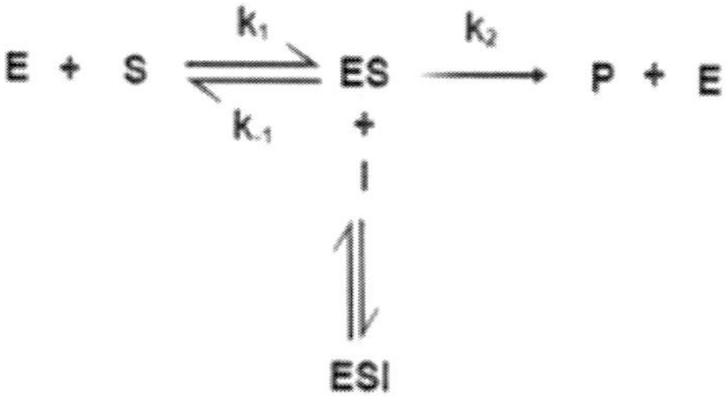

$$E + S \underset{k_{-1}}{\overset{k_1}{\rightleftharpoons}} ES \xrightarrow{k_2} P + E$$

$$ES + I \rightleftharpoons ESI$$

Figure 5. Schematic representation of a competitive inhibition.

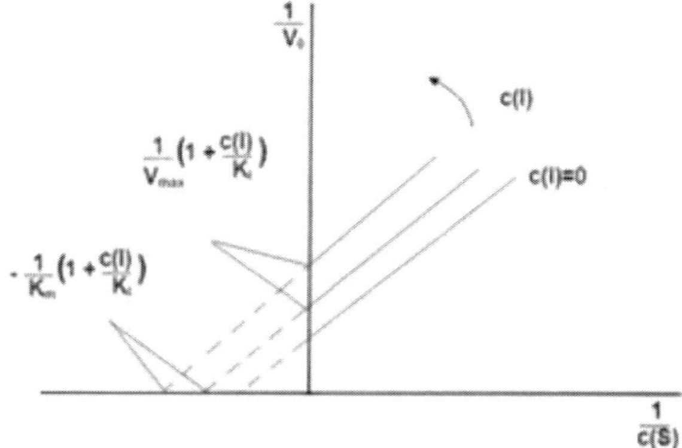

Figure 6. Lineweaver-Burk diagram for a competitive inhibition.

A Competitive Inhibition

The characteristic of this type of inhibition is that the inhibitor reversibly binds to the enzyme-substrate complex, creating an inactive enzyme-substrate-inhibitor complex that leads to the formation of the product (Figure 5).

In this case, the Lineweaver-Burk diagram takes another form (Figure 6):

$$\frac{1}{v_0} = \frac{K_M}{V_{max}}\frac{1}{[S]} + \frac{1}{V_{max}}\left(1 + \frac{[I]}{K_I}\right) \quad (9)$$

With this type of inhibition, a decrease in the Vmax value is observed, but there is no effect on the K_M constant. An example of an a competitive inhibitor is the binding of epristeride to the enzyme steroid 5a-reductase (Copeland and Anderson, 2002). This enzyme binds the cofactor NADPH to the active site and then binds the male sex hormone testosterone to form an enzyme-NADPH-testosterone ternary complex.

Enzyme Inhibition of Drug

Drugs can act on enzymes as competitive, a competitive or non-competitive inhibitors of catalase, irreversible or reversible. Another way drugs act on enzymes is to act as false substrates. Then the drug molecules undergo a chemical transformation and become inappropriate products that block the normal development of the metabolic pathway (Rang, Riter, Flauer and Henderson, 2005).

The choice of enzymes as drug targets is not only due to their catalytic activity but also since enzymes are very susceptible to inhibition by small molecules similar to drugs. Today, there is a vast number of drugs in clinical use that function as competitive enzyme inhibitors (Rashtbari, Dehghan, Yekta, and Jouyban, 2017). Mechanism and structure-based drug design refer to the ability to design competitive enzyme inhibitors based on knowledge of the catalytic reaction mechanism, using substrates, transition states, or mimicking product states as starting points in design (Copeland and Anderson, 2001). A good example of this strategy is discovering the antihypertensive drugs captopril and enalapril (Navia and Murcko, 1992; Wlodawer and Vondrasek, 1998).

Effective angiotensin-converting enzyme (ACE) inhibitors, such as the antihypertensive drugs captopril and enalapril, act by chelating the zinc atom and thus disrupt the catalytic component of the enzyme's active site, although they do not occupy the entire volume of the active site cleft. Due to the influence of drugs on the catalytic activity of enzymes, the drug discovery and design approach are fundamental. However, it is essential to recognize that the active site of an enzyme is not the only one on the enzyme molecule that can be a suitable target for drug interactions (Copeland and Anderson, 2002). Antihistamines also block the active site of catalase. Some of the actions of H1-receptor antagonists do not block H1-receptors. Still, they result from antagonistic effects at the level of other receptors (5-hydroxytryptamine/5-HT, α1-adrenergic receptors and muscarinic receptors) in the peripheral and central nervous systems. Some H1-receptor antagonists exert adverse effects on the central nervous system or may be clinically more useful than peripheral effects. An example of this is quite strong sedatives. Certain H1-receptor antagonists exhibit an antimuscarinic effect even though they have a higher affinity for histamine. Non-sedating antihistamines, such as loratadine and cetirizine, have been developed, but they exhibit adverse events such as heart rhythm disorders. Among the H1-receptor antagonists, some exert a weak blockade of α1-adrenergic receptors. The clinical use of histamine H1-receptor antagonists has been proven in allergic reactions and allergic rhinitis, e.g., hay fever, urticaria, insect bite, and drug hypersensitivity. In therapy, drugs without sedative or antagonistic effects on muscarinic receptors have a higher preference (Assanasen and Naclerio, 2002). Lastly, H1-receptor antagonists are used as antiemetics in preventing motion sickness or other causes of nausea, especially when vertigo is also present (Rang, Riter, Flauer and Henderson, 2005).

Boroxins are another example of enzyme inhibitors. Thanks to their unique electronic structure, boroxines are currently used as potential enzyme modifiers and therapeutics. Many studies have shown that dipotassium trioxohydroxytetrafluorotriborate, $K_2[B_3O_3F_4OH]$, stands out as an enzyme inhibitor that can be used as a therapy in cancer patients as well as in the prevention and treatment of benign and malignant skin changes, which manifest in the form of tumors. If growths on the skin are repeatedly brought into contact with very small amounts of $K_2[B_3O_3F_4OH]$, their withdrawal occurs. Previous research explored the kinetic parameters and mechanism of inhibition of $K_2[B_3O_3F_4OH]$ on the enzyme catalase, given that a potential drug in the cell can affect the change in enzyme activity. It was discovered that the presence of $K_2[B_3O_3F_4OH]$ reduces catalase activity and increases hydrogen peroxide production, which destroys tumor cells. Although the mechanism of action of $K_2[B_3O_3F_4OH]$ on tumor cells is not fully explained, it has been proven that this substance specifically recognizes tumor cells and destroys them while healthy cells remain intact (Islamovic, Galic and Milos, 2013).

Methods

Different methods can be used to determine catalase inhibition, and this chapter will describe spectrophotometric and electrochemical methods. A spectrophotometric method was used to analyze the effect of a drug on catalase activity *in vitro*. In the paper by Hasković et al. 2021, a UV/Vis monochromatic spectrophotometer was measured at a wavelength of 405 nm., using the method described by Góth L. (Góth L, 1991). The effects of six different concentrations of hydrogen peroxide substrates were tested with buffer, catalase, loratadine, and an incubation period at 37°C for one minute. The enzymatic reaction was stopped with a solution of ammonium heptamolybdate tetrahydrate (Hasković, 2021.) A potentiostat/galvanostat PAR 263A with a conventional three-electrode system was used for the electrochemical measurements. A saturated Ag/AgCl electrode was used as the reference electrode, a Pt electrode was used as the counter electrode, and a GC electrode as the working electrode. An amperometric biosensor for the determination of H_2O_2 was formed by the immobilized enzyme catalase (CAT) captured in a Nafion layer on the surface of the GC electrode (Herenda et al., 2018).

Electrochemical methods of chronoamperometry were used for the measurements. A chronoamperometric technique was used to determine the kinetic parameters, i.e., the Michaelis-Menten constant (K_M) and the maximum current value at which the substrate is saturated with the substrate (I_{max}), which is equivalent to the maximum reaction rate (V_{max}) (Adeyoju, 1995). Chronoamperometric measurements were carried out in the electrochemical cell containing 25 mL of the phosphate buffer solution with a constant potential of 0.9 V applied to the working electrode and constant mixing at 400 rpm.

Result and Discussion

The effect of six different concentrations of hydrogen peroxide substrates was tested (0.13; 0.26; 0.4, 0.53; 0.66, 1.0 mM), with buffer, catalase, loratadine (Hasković, 2021). The Lineweaver-Burke diagram shown in Figure 7 shows that a competitive type of inhibition is present in the enzymatic reactions performed, and the K_i value with loratadine is 6.0 mM.

However, the resulting V_{max} values change insignificantly, while the K_M values decrease with increasing inhibitor concentration (Table 1).

The above indicates that a partial competitive type of inhibition is present in the *in vitro* enzyme reactions. Loratadine belongs to the second generation of H1-antihistamines and has a therapeutic effect in alleviating inflammatory processes. In other words, it alleviates the symptoms of allergic rhinitis, conjunctivitis, and urticaria. *In vitro,* loratadine exhibits anti-allergic activity unrelated to H1-histamine receptors. Also, loratadine acts as a partial inhibitor in the antigen-mediated release of leukotrienes from human bronchi (Berthon, Taudou, Combettes, et al. 1994). Islamovic et al. examined the influence of boroxine on enzyme activity using the gasometric method. The measurement of released oxygen as a function of the time was done in a slightly modified apparatus described by Schubert at a temperature of 37°C with constant stirring. The total reaction volume of 25 mL, containing a buffer solution with a constant value of catalase and the appropriate amount of $K_2[B_3O_3F_4OH]$, was thermostated, and then a proper substrate amount of H_2O_2 was added. From the experimental data, the initial velocity Vo was calculated for the appropriate substrate concentration and $K_2[B_3O_3F_4OH]$. Lineweaver–Burk plots at different fixed concentrations of $K_2[B_3O_3F_4OH]$ were linear and plotted of straight lines that intersect the x-axis at the same point ($-1/K_M = -0.05$) (Islamovic, Galic and Milos, 2013).

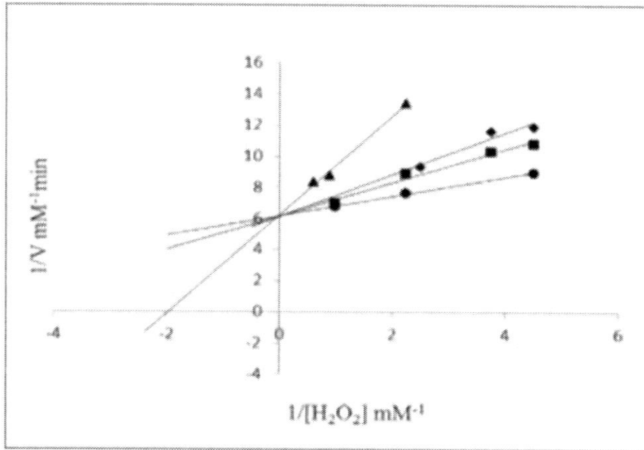

Figure 7. Lineweaver-Burk diagram for the determination of V_{max} and K_M in the absence and presence of different concentrations of loratadine: ▲0 g/L; ♦0.06 g/L; ■ 0.2 g/L; ● 0.26 g/L.

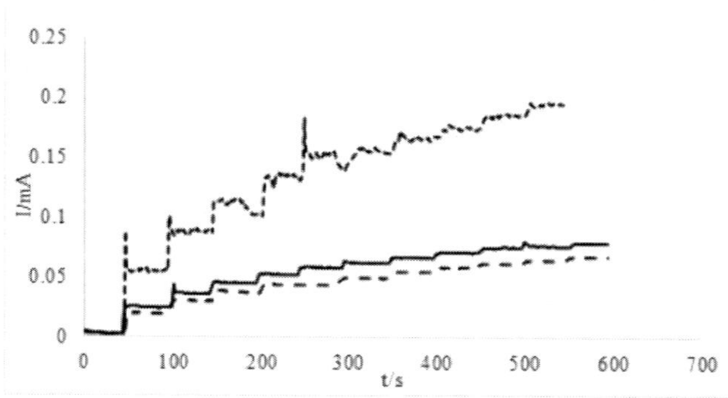

Figure 8. Chronoamperogram without the presence and with the presence of different concentrations of diclofenac: ---- 0 mM; — 0.112 mM; - - - 0.168 mM.

Table 1. Values of K_m and V_{max} with and without the presence of different concentrations of loratadine

[loratadine] g/L	V_{max} (mMmin^{-1})	K_M (mM)
0	0.1555	0.4742
0.06	0.1634	0.2198
0.2	0.1677	0.1854
0.26	0.1617	0.1002

In the absence of an inhibitor, the kinetic studies on the activity indicate that the enzyme has a K_M value of 20 mM and V_{max} value of 0.22 mmoles $L^{-1}s^{-1}$, which is in good correlation with the common value for K_M and V_{max} of catalase observed in the literature. In the presence of $K_2[B_3O_3F_4OH]$ from the straight lines intersection of the x-axis, the K_M for H_2O_2 was found to be constant and equal to 20 mM, and V_{max} were variable 0.18, 0.13 and 0.12 mmoles $L^{-1}s^{-1}$, respecting concentration of $K_2[B_3O_3F_4OH]$. Accordingly, although unusual in practice, it seems that $K_2[B_3O_3F_4OH]$ is a classical non-competitive inhibitor, and non-competitive inhibition occurs. Therefore, local application of $K_2[B_3O_3F_4OH]$ containing cream or its intratumor injection of millimolar concentrations could significantly reduce catalase activity and increase the concentration of H_2O_2, thus producing beneficial effects in tumor tissue alone (Islamovic, Galic and Milos, 2013). Electrochemical measurements of the effect of diclofenac on catalase activity were performed using the chronoamperometry method (Štitkovac, 2020). Measurements were made without the presence and with the presence of different concentrations of diclofenac: 0.056 mM; 0.112 mM; 0.168 mM; 0.223 mM, and 0.277 mM for a fixed concentration of immobilized catalase of 7.01 x 10^{-3} g/mL, with the addition of different substrate concentrations at intervals. 100 μL of hydrogen peroxide was added every 50 s (Figure 8).

The maximum speed of the reaction decreases, and the Michaelis-Menten constant remains almost unchanged for different concentrations of diclofenac and has a value of $K_M \approx 9$ mM, which indicates that diclofenac binds reversibly to the enzyme at a site that is not specific for the substrate, and thus reduces the catalytic activity of enzymes. Diclofenac, an anti-inflammatory drug used in painful, inflammatory, rheumatic and sometimes non-rheumatic conditions, is a non-competitive inhibitor of catalase. It has also been proven that its action is based on the inhibition of cyclooxygenase (COX) activity and the consequent formation of pro-inflammatory mediators such as prostaglandins (PGs) and thromboxanes (Naveed and Qamar, 2014). Enzyme inhibition means reducing or blocking the action of an enzyme on a specific part of the active site of the enzyme by a specific substrate or enzyme inhibitor.

The action of enzyme inhibitors in drug detection has become a fundamental approach in pharmacology in any pharmaceutical industry, university research laboratory, or drug research center (Copeland, 2005). Corticosteroid treatment is one of the most commonly used and effective treatments for various inflammatory and autoimmune disorders. The Michaelis-Menten kinetic model examines the inhibitory effect of different concentrations of betamethasone dipropionate in the range of hydrogen peroxide substrate concentrations (1.5-2.5 mM) on enzyme activity, used in the work of the authors Herenda et al., 2020 (Figure 9).

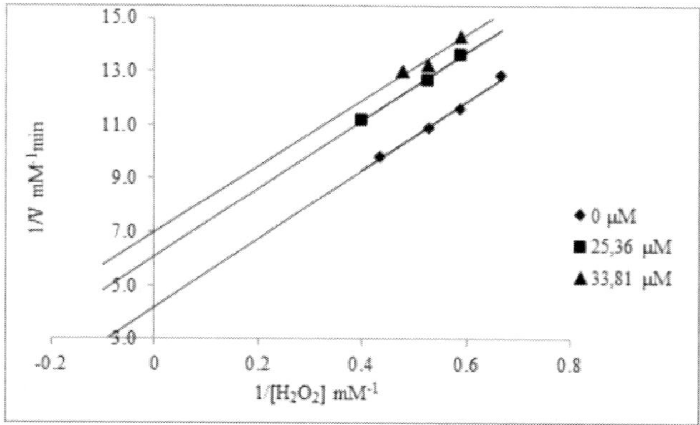

Figure 9. Lineweaver - Burk diagram for determination of V_{max} and K_M without the presence and with the presence of different concentrations of betamethasone dipropionate.

Based on the graphical presentation of the effects of betamethasone dipropionate, it is evident that it is an a competitive type of enzyme inhibition, which is indicated by parallel directions and constant slopes for each concentration of inhibitor. Inhibitors of this modality require the prior formation of ES binding and inhibition complexes. Therefore, these inhibitors affect the catalysis steps after the initial non-covalent binding of the substrate to the active site of the enzyme. The betamethasone intermediate occurs when stereospecific hydride electron transfer occurs from the substrate to the glucosteroid double bond. This enolate intermediate is stabilized by interaction with an acid group within the enzyme active site. Then comes the donation of protons from the active site to enolate carbon, forming the reaction product (Schramm, 2005; Copeland and Anderson, 2001).

Conclusion

Research has shown that the enzyme catalase follows the Michaelis-Menten kinetic model. Due to their structure, many drugs can bind differently to the site on catalase and thus reduce the formation of free radicals, i.e., the development of many diseases in the body. Physiologically, catalysts are critical, as they greatly facilitate the speed of common chemical reactions. Without catalase, essential reactions would take days or months, and toxic H_2O_2 would accumulate quickly in the body. Reactive oxygen species (ROS) are toxic and function as signaling molecules. Resistance of tumor cells against intercellular ROS signaling depends on interference through catalase expression on the cell membrane. Intercellular ROS signaling of tumor cells can be restored when H_2O_2 is supplied, and catalase is inhibited.

Disclaimer

None

References

Adeyoju, O. O. (1995). "*Development of horseradish peroxidase and tyrosinase-based organic-phase biosensors.*" PhD diss., Dublin City University.

Assanasen, P. and Naclerio, R. M. (2002). Antiallergic anti-inflammatory effects of H1-antihistamines in humans. *Clin. Allergy Immunol.*, 17: 101-39.

Berthon, B., Taudou, G., Combettes, L., et al. (1994). *In vitro* inhibition, by loratadine and descarboxyethoxyloratadine, of histamine release from human basophils, and of histamine release and intracellular calcium fluxes in rat basophilic leukemia cells (RBL-2H3). *Biochem. Pharmacol.*, 47(5): 789-94.

Bisswanger, H. (2008). *Enzyme kinetics. Principles and methods.* Weinheim: Wiley-VCH Verlag GmbH & Co. KGAa.

Bugg, T. D. H. (2012). *Introduction to Enzyme and Coenzyme Chemistry*, 3rd Edition. New Jersey: John Wiley & Sons.

Christopoulos, A. (2002). Allosteric binding sites on cell-surface receptors: novel targets for drug discovery. *Nature Rev. Drug Discov.* 1: 198-209.

Copeland, R. A., and Anderson, P. S. (2001), *Enzymes and Enzyme Inhibitors in Textbook of Drug Design and Discovery*, 3rd ed., P. Krogsgaard-Larsen, T. Liljefors, and U. Madsen, eds., Taylor and Francis, New York, 328-363.

Copeland, R. A. *Evaluation of Enzyme Inhibitors in Drug Discovery*, Hoboken: John Wiley & Sons, Inc., 2005.

Cornish-Bowden, A. (1995). *Fundamentals of Enzyme Kinetics*, Portland Press, London.

Djordjević, V. B. (2004). Free Radicals in Cell Biology. *International Review of Cytology*, 237, 55-89.

Fontes, R., Ribeiro, J. M. and Sillero, A. (2000). Inhibition and activation of enzymes. The effect of a modifier on the reaction rate and on kinetic parameters. *Acta Biochim. Pol.*, 47(1): 233-57.

Furman, P. A., Painter, G. R. and Anderson, K. S. (2000). An Analysis of the Catalytic Cycle of HIV-1 Reverse Transcriptase Opportunities for Chemotherapeutic Intervention Based on Enzyme Inhibition. *Current Pharmaceutical Design*, 6(5): 547-567.

Góth, L. (1991). A simple method for determination of serum catalase activity and revision of reference range. *Clinica Chemica Acta*, 196 (2-3): 143-151.

Hasković, E., Herenda, S., Halilović, Z., Unčanin, S., Hasković, D., & Deljkić, E. (2021). Investigation of the Effect of Loratadine and Calcium Ions on Oxidoreductase Activity of Catalase Enzyme. *Current Enzyme Inhibition*, 17, 1-7.

Herenda, S., Crnkić, M., Klepo, L., Hasković, E. et al. (2021). Influence of betamethasone dipropionate and prednisone on enzyme kinetics. *Farmacia*, 69(6):1060-1065.

Herenda, S., Ostojić, J., Hasković, E., Hasković, D. et al. (2018). Electrochemical Investigation of the Influence of $K_2[B_3O_3F_4OH]$ on the Activity of Immobilized Superoxide Dismutase. *Int. J. Electrochem. Sci.*, 13:3279-3287.

Islamovic, S., Galic, B. and Milos, M. (2014). A study of the inhibition of catalase by dipotassium trioxohydroxytetrafluorotriborate $K_2[B_3O_3F_4OH]$. *J. Enz. Inh. and Med. Chem.*, 29:744.

Karlson, P. (1993). *Biokemija* (VIII izdanje) [*Biochemistry* (VIII edition)]. Zagreb: Školska knjiga.

Marangoni, A. G. 2003. *Enzyme Kinetics. A Modern Aproach*. New Jersey: John Wiley & Sons.

Miholjčić, M., Jadrid, S. and Winterhalter-Jadrid, M. (1988). *Biohemija*. Zavod za udžbenike i nastavna sredstva, Sarajevo [*Biochemistry*. Institute for textbooks and teaching aids, Sarajevo].

Milton, N. G. N. (2008). Homocysteine inhibits hydrogen peroxide breakdown by catalase. *The Open Enzyme Inhibition Journal*, 1(1), 34-41.

Naveed, S. and Qamar, F. (2014). UV spectrophotometric assay of Diclofenac sodium available brands. *Journal of Innovation in Pharmaceuticals and Biological Science*, 1(3):92-96.

Navia, M. A. and Murcko, M. A. (1992). Use of structural information in drug design. *Curr. Opin. Struct. Biol.* 2: 202-210.

Navia, M. A., Fitzgerald, P. M., McKeever, B. M., Leu, C. T. et al. (1989). Three-dimensional structure of aspartyl protease from human immunodeficiency virus HIV-1. *Nature*, 337: 615-620.

Palmer, T. and Bonner, P. L. (2007). *Enzymes: Biochemistry, Biotechnology and Clinical Chemistry*, Cambridge: Woodhead Publishing Limited.

Purich, D. L. and Allison, R. D. (2002). *The Enzyme Reference: A Comprehensive Guidebook to Enzyme Nomenclature, Reactions, and Methods*, Boston: Academic Press.

Rang, H. P., Riter, J. M., Flauer, R. J. and Henderson, G. (2005). *Farmakologija* [*Pharmacology*]. Beograd.

Rashtbari, S., Dehghan, G., Yekta, R. and Jouyban, A. (2017). Investigation of the binding mechanism and inhibition of bovine liver catalase by quercetin: Multi-spectroscopic and computational study. *Bioimpacts*, 7(3): 147-53.

Schramm, V. L. (2005). Enzymatic transition states and transition state analogs, *Curr. Opin. Struct. Biol.*, 15:604-613.

Schubert, W. M. (1949). The aromatic elimination reaction. I. Mechanism of the decarboxylation of mesitoic acid. *J. Am. Chem. Soc.*, 71: 2639-44.

Štitkovac, N. (2020). "Elektrohemijsko određivanje uticaja antiinflamatornog lijeka na aktivnost katalaze" ["*Electrochemical determination of the effect of an anti-inflammatory drug on catalase activity*"]. Master thesisi. University of Sarajevo.

Wlodawer, A. and Vondrasek, J. (1998). Inhibitors of HIV-1 protease: a major success of structure-assisted drug design. *An. Rev. Biophys. Biomol. Struct.*, 27: 249-284.

Index

R

S

T